T0077581

GMAT

GMAT

Mathivanan Palraj

PARTRIDGE

ISBN: Softcover 978-1-4828-7333-7

Print information available on the last page.

To order additional copies of this book, contact
Partridge India
000 800 10062 62
orders.india@partridgepublishing.com

www.partridgepublishing.com/india

Part-I: Problem Solving Techniques for Top Score

CONTENTS

1. SIMPLIFICATION

An obstacle

Many students do not get the required cut-off in quant section. Many reasons can be attributed to this problem. One reason is that they take a lot of time to simplify the arithmetic operations involved in a certain problem. This difficulty can be overcome if they practice regularly. Your confidence will seriously be undermined without this basic skill. You might even commit a silly mistake (addition, subtraction, multiplication or division) in one simplification and fail to get the cut-off. Your life-long dream is shattered with that one silly mistake. Would you dare to take such risk? One of my students committed three silly mistakes in three different problems and got all the three wrong in the last test. If she got all the three questions correct, she would have cleared the cut-off.

Improving the skill

Study the multiplication table up to 20. Group study will improve you a lot. Ask your friend, what is 12x18? Fix a time limit, if your friend is not answering within the time you will tell the answer and check it too! Now it is your friend's turn. Repeat this for ten minutes a day and see the progress. You will wonder! Gradually take bigger numbers. You will be able to do it.

One of my students would go to the highways and add the numbers on the number plate of a certain vehicle before the next vehicle to pass by. You can also do multiplication. Does that sound weird? No, it does not. It helps. It is fun mathematics. **Warning:** Take a safe place.

Some of my students used to play a game with playing cards. Everyone will be distributed with one card. The person who is having the highest number has to add or multiply the numbers of all cards distributed. If the addition or multiplication is wrong, the person will be awarded a penalty. The difficulty level will be higher when there are more participants.

When you are alone and cannot go out, do this:
Take two two-digit numbers and multiply. You just have to write the answer!
12x18 = 216

Rule: The unit digit 6 is obtained by multiplying the first digits of the two numbers, 2x8=16. Write 6 and carry 1.
The ten's digit 1 is obtained by adding the product of inner digits, product of outer digits and carry forward of 16, that is 1. (2x1 + 1x8 +1) = 11.
The hundred's digit 2 is obtained by adding the product of second digits and the carry forward.
1x1 +1 = 2

Another method, which is specific to this particular kind of pair

Rule: if the sum of unit digits is 10 and the second digits are the same, the first two digits are obtained by multiplying the 1st digit and 1st digit (2x8=16). The next numbers are obtained by adding 1 to the 2nd digit and then multiplying it by the 2nd digit (1+1=2; 2x1=2). Try 24x26=
4x6=24
3x2=6
Therefore, the answer is 624.

This is more difficult than the conventional method. In the conventional method, you write down everything and do not need to memorise products and additions. Here, you have to multiply, memorise the result, then add something, and then again memorise something...

Why should I recommend this method?

You improve your working memory. You improve your intelligence. Gradually you will do all your calculations mentally! I bet you would do 50% more problems in the test.

You can also do the product of two three-digit numbers once you mastered the product of two-digit numbers.
123x456 = 56088

Rule: The unit digit 8 is obtained by multiplying 6x3=18.
The ten's digit 8 is obtained by adding the products of the second digit of the first number and the first digit of the second number, the first digit of the first number and the second digit of the second number, and the carry forward. 5x3+6x2+1=28.
The hundred's digit 0 is obtained by adding all the products of the first digit of the first number and the third digit of the second number, the second digit of the first number and the second digit of the second number, the third digit of the first number and the first digit of the second number, and the carry forward. 6x1+5x2+4x3+2.

The thousand's digit 6 is obtained by adding all the products of the third digit of the first number and the second digit of the second number, the second digit of the first number and the third digit of the second number, and the carry forward. 4x2+5x1+3.
The ten thousand's digit 5 is obtained by adding the product of the third digit of the first number and the third digit of the second number, and the carry forward.4x1+1.
Hope you got the logic.

Some multiplications you can easily memorise

Squaring the numbers which end in 5

15x15=225; 25x25=625; 35x35=1225; 45x45=2025; 55x55=3025; 65x65=4225

Rule: All the results end with 25. It is obtained by multiplying the unit digit and unit digit. The other numbers are obtained by adding 1 to the 2nd digit and then multiplying by the 2nd digit (for 15, (1+1)1=2; for 25, (2+1)2=6; for 35, (3+1)3=12 and so on). Therefore, 85x85=
5x5=25
9x8=72
7225.

105x105=
5x5=25
11x10=110
Therefore, the answer is 11025.

Multiplying any number with 11
11x11=121
12x11=132
13x11=143

14x11=154
19x11=209

Rule: write the unit digit as it is. The tens digit is obtained by adding the first two digits. The hundred's digit is obtained by adding carry forward and the 2nd digit. For example,
23x11=
Unit's digit is 3
Ten's digit is 2+3=5
Hundred's digit is 2 and no carry forward.
Therefore, the answer is 253.
47x11
Unit's digit is 7
Ten's digit is 1 (carry forward 1, (4+7=11))
Hundred's digit is 4+1=5
Therefore, the answer is 517.

179x11
Unit's digit is 9
Ten's digit is 6(9+7=16) and carry forward 1 (adding the first two digits)
Hundred's digit is 1+7+1=9 (adding the next two digits and carry forward)
The last digit is 1
Therefore, the answer is 1969.

BASIC SIMPLIFICATIONS

EG1: $4\frac{1}{2}+3\frac{1}{6}=?$

Solution:
Add the whole numbers first; 4+3=7
Add the fractions next; the LCM of 2 and 6 is 6
Convert the smaller denominator into LCM by multiplying the numerator and denominator by 3

$\frac{3}{6}+\frac{1}{6}=\frac{4}{6}=\frac{2}{3}$

Therefore, the answer is $7\frac{2}{3}$.

EG2: $4\frac{1}{2}+3\frac{3}{4}=?$

Solution:

$\frac{2}{4}+\frac{3}{4}=\frac{5}{4}=1\frac{1}{4}$

Therefore, the answer is $8\frac{1}{4}$.

EG3: $4\frac{1}{2}-3\frac{1}{6}=?$

Solution:

First subtract the whole numbers; 4-3=1

Next, subtract the fractions.

$\frac{3}{6}-\frac{1}{6}=\frac{2}{6}=\frac{1}{3}$

Therefore, the answer is $1+\frac{1}{3}=1\frac{1}{3}$.

EG4: $4\frac{1}{6}-3\frac{1}{2}=?$

Solution:

4-3=1

$\frac{1}{6}-\frac{3}{6}=-\frac{2}{6}=-\frac{1}{3}$

Therefore, the answer is $1-\frac{1}{3}=\frac{1}{1}-\frac{1}{3}=\frac{3}{3}-\frac{1}{3}=\frac{2}{3}$.

EG5: $4\frac{1}{6}\times 3\frac{1}{2}=?$

Solution:

Method1:

Convert the mixed fractions into improper fractions and multiply

$\frac{25}{6}\times\frac{7}{2}=\frac{175}{12}=14\frac{7}{12}$.

Method2: it improves your mental calculations

First, multiply the fractions, $\frac{1}{6}\times\frac{1}{2}=\frac{1}{12}$

Next, multiply the 1st fraction and 2nd whole number; multiply the 2nd fraction and 1st whole number and add.

$4\times\frac{1}{2}+3\times\frac{1}{6}=\frac{4}{2}+\frac{3}{6}=2\frac{1}{2}$

Next, multiply the whole numbers; 4x3=12

Add all the results.

$$12+2+\frac{1}{2}+\frac{1}{12}=14\frac{7}{12}$$

EG6: $4\frac{1}{6}\div 3\frac{1}{2}=?$

Solution:

First, convert the mixed fractions into improper fractions.

$$4\frac{1}{6}=\frac{6\times 4+1}{6}=\frac{25}{6};3\frac{1}{2}=\frac{2\times 3+1}{2}=\frac{7}{2}$$

$$\frac{25}{6}\div\frac{7}{2}$$

Next, take the reciprocal of the second fraction and multiply with the first.

$\frac{25}{\not{6}_3}\times\frac{\not{2}}{7}=\frac{25}{21}=1\frac{4}{21}$ (While multiplying you can cancel the common factors in the numerator and denominator)

EG7:108÷36 of $\frac{1}{4}$=?

Solution:

You have to apply **BODMAS** rule. First, simplify 'of' then 'division'. 'Of' is nothing but multiplication.

$36\text{x}\frac{1}{4}=9$

108÷9=12

EG8: If $\frac{1}{2}+\frac{1}{3}+\frac{1}{x}=1$ then what is the value of x?

Solution:

First, add $\frac{1}{2}+\frac{1}{3}=\frac{5}{6}$

Next, take this value to the right hand side.

$\frac{1}{x}=1-\frac{5}{6}$(When certain value goes from one side to the other side, its sign changes; e.g. from +ve to –ve)

$$\frac{1}{x}=\frac{6\times 1-5}{6}=\frac{1}{6}$$

Therefore, the value of x = 6

EG9: $\frac{1}{2}\times\frac{1}{3}\times\frac{1}{x}=\frac{5}{2}$; what is the value of x?

Solution:

First, multiply $\frac{1}{2}\times\frac{1}{3}=\frac{1\times1}{2\times3}=\frac{1}{6}$

Next, take this value to the right hand side.

$\frac{1}{x}=\frac{5}{2}\times\frac{6}{1}=\frac{15}{1}$(if it is multiplication, first take the reciprocal and then multiply)

Therefore, the value of x $=\frac{1}{15}$

EG10: Find the LCM of 12, 18 and 20

Solution:

Out of the three numbers 20 is the highest number. Therefore, take the multiples of 20 and check whether it is divisible by the other two numbers. Since one of the numbers is ending with '0', the LCM will also end with '0'. The multiples of 18, which are ending with '0', are 90, 180, and 270 and so on. Out of these numbers only 180 is divisible by all the three numbers. The answer is 180. (**Here, I used an unconventional method.)**

PRIME FACTOR METHOD

$12 = 2^2\times3$

$18 = 2\times3^2$

$20 = 2^2\times5$

To find the LCM, take all the prime factors with highest power and then multiply all.

LCM = $2^2\times3^2\times5$ = 180

To find the HCF, take all the common prime factors with lowest power and then multiply all. HCF = 2 (2 is the only common factor and its lowest power is 1)

Hope, you have learned the basic arithmetic operations. The first hurdle is over. The next step is learning the concepts. In the beginning of each chapter the concepts for the particular chapter are given. Read thoroughly and build your concepts. Then do the solved examples. While doing the solved examples try to solve yourselves without referring the solution. If you cannot do it, you shall see the solution. Then go to the exercise, which will take you to the next level. The exercise is designed in such a way that you will have to apply the concepts in certain difficult problems. This will equip you to tackle the higher-level problems. Wherever it is necessary, I have provided you with problem solving techniques, which is a unique feature of this book. Learn the techniques and it will enable you tackle many difficult problems.

You have to develop your application mind. This particular trait is invariably observed among the students who top-scored in the test. It is not just another college examination. It is designed to test your suitability for the profession. It tests your knowledge. It tests whether you can apply this knowledge in certain situation. It tests whether you apply the right technique at right place. Will you be able to apply your brain under pressure? Some are very knowledgeable. However, they cannot perform under pressure! This particular negative trait is not uncommon. Get rid of this attitude and you will be successful. Your score will be better when you are steady and composed.

2. NUMBERS

Natural numbers: 1, 2, 3, 4, ..., n

Whole numbers: 0, 1, 2, 3, 4, ...,n

Integers:.......-4, -3,-2, -1, 0, 1, 2, 3, 4, ...

Rational numbers: numbers, which are in the form of p/q where p & q are integers and $q \neq 0$

Irrational number: all real numbers, which are not in the form of p/q, are irrational. They are non-terminating and non-recurring. π , e are examples of irrational numbers.

Complex numbers: numbers in the form of a + ib, where a and b are real and ib is imaginary part, are called complex numbers. $i^2 = -1$

Prime number: is a positive integer greater than 1, which is divisible only by the same number and by 1.

Composite number: all positive integers which are greater than 1 and which are not prime numbers

Co-primes: two positive integers, which are greater than 1, are co-primes if they do not have any common factor between them except 1. 3 and 4 are co-primes, though 4 is not a prime number.

Factor: 8 = 2x4, 8 is a composite number, which is divisible by 2 as well as 4. 2 and 4 are called factors. 2 is a prime factor and 4 is simply a factor because 4 is not a prime number. Note that any composite number can be written as a product of prime factors.

8 = 2x2x2
9 = 3x3
15 = 3x5

Number of factors:
If N is a composite number and $N = a^p \times b^q \times c^r ...$ where a, b and c are prime factors and p, q and r are positive integers then the number of factors of N is equal to (p+1)(q+1)(r+1)....

Sum of all the factors:
If N is a composite number and $N = a^p \times b^q \times c^r ....$ then the sum of all factors of N is equal to $\frac{a^{p+1}-1}{a-1} \times \frac{b^{q+1}-1}{b-1} \times \frac{c^{r+1}-1}{c-1} \times ......$

DIVISIBILITY CONDITIONS

All even numbers are divisible by 2
If the sum of the digits of a number is divisible by 3 then the number is divisible by 3

If the last two digits that is on the right side is divisible by 4 then the number is divisible by 4

If the unit digit is either 0 or 5, then the number if divisible by

If a number is divisible by 2 as well as 3, then the number is divisible by 6

If the last 3 digits of a number are divisible by 8 then the number is divisible by 8

If the sum of the digits of a number is divisible by 9 then the number is divisible by 9

If the difference between the sum of odd digits and even digits of a number is either 0 or multiple of 11 then the number is divisible by 11

If a number is divisible by 3 as well as 4 then the number is divisible by 12. Note that you cannot check for 2 and 6 because they are not co-primes, whereas 3 and 4 are co-primes.

Divisibility condition for 7, 13, 17, 19, 23, and 29

Divisibility condition for the above prime numbers is a bit hard to memorize. However, if you can master the method it will be useful.

Fortunately, there is a general rule, which you can apply to all these numbers.
Step 1
Find a multiple of the number which is 1 less than or 1 greater than a number which is a multiple of 10. For example, 21 is a multiple of 7 and is 1 greater than 20 which is a multiple of 10.

Step 2
21 is a multiple of 7 and is equal to (2x10+1). You have to remember two things: a) the key number is 2 and the sign is negative because you have to reduce 1 to get the number 20. Therefore, the key for the number 7 is -2.

39 is a multiple of 13 and is equal to (4x10-1). Therefore, the key for the number 13 is +4 and it is positive because you have to add 1 to get the number 40, which is a multiple of 10.

How to check whether a number is divisible by 7

Consider 553. The simple technique is eliminating the digits one by one starting from the unit digit. The key for 7 is -2. Multiply the unit digit with the key and add the result to the remaining number 55. When you do that, you get 55-6=49 (now you have eliminated one digit). 49 is divisible by 7. Therefore, 553 is also divisible by 7.

Thus, you can consider any number and check for the divisibility of 7, 13, 17, 19, 23, or 29.

Certain results, which are useful

$(x^n - a^n)$ is divisible by $(x - a)$ for all values of n
$(x^n - a^n)$ is divisible by $(x + a)$ for all even values of n
$(x^n + a^n)$ is divisible by $(x + a)$ for all odd values of n

BASIC FORMULAE
$(a+b)^2 = a^2+b^2+2ab$
$(a-b)^2 = a^2+b^2-2ab$
$(a^2-b^2) = (a+b)(a-b)$
$(a+b+c)^2 = a^2+b^2+c^2+2(ab+bc+ca)$
$(a+b)^3 = a^3+b^3+3ab(a+b)$
$(a-b)^3 = a^3-b^3-3ab(a-b)$
$a^3+b^3 = (a+b)(a^2-ab+b^2)$
$a^3-b^3 = (a-b)(a^2+ab+b^2)$
$(a^3+b^3+c^3-3abc) = (a+b+c)(a^2+b^2+c^2-ab-bc-ca)$

If $a+b+c = 0$, then $a^3+b^3+c^3 = 3abc$

HIGHEST COMMON FACTOR

HCF of two or more numbers is the greatest number that divides exactly each of them.

LEAST COMMON MULTIPLE

LCM of two or more numbers is the lowest multiple that can be exactly divided by each of the numbers.

Product of two numbers = LCM of the two numbers x HCF of the two numbers.

LCM of fractions = LCM of numerators / HCF of denominators.

HCF of fractions = HCF of numerators / LCM of denominators.

SOLVED EXAMPLES

1. The LCM and HCF of two numbers are 84 and 21 respectively. If the ratio of the two numbers is, 1:4 then find the largest number.

Solution:

Ratio of the two numbers is 1:4
Let the two numbers are x and 4x
LCM of x and 4x is 4x => x = 84/4 = 21
Therefore, the numbers are 21 and 84
The largest number is 84
##In the above problem, you can find the answer even if the ratio is not given. Let the two numbers are 21x and 21y. LCM is 21xy. Therefore, xy=4. There are two possibilities: (1, 4) or (2, 2). The second set is obviously wrong.

2. If one-third of a number is 3 more than one-fourth of the number then what is the number.

Solution:

Let the number be x. Therefore, x/3 = x/4 + 3
x/3 – x/4 = 3; x/12 = 3 => x = 36
##You can also solve this problem by assuming **a convenient number**. The convenient number here is the LCM of 3 and 4 which are the denominators of the ratios given in the problem, one-third and one-fourth. One-third of 12 is 4 and one-fourth of 12 is 3. The difference is 1. If the difference is 3 then the required number is 3x12=36

3. If one-eighth of a pencil is black, half of the remaining is yellow and the remaining three and a half cm is blue, then the total length of the pencil is

Solution:

Let the length of the pencil be x
Therefore, x/8 = black
x – x/8 = 7x/8 is remaining
½ [7x/8] = 7x/16 = yellow
Hence x/8 +7x/16 +7/2 = x
(2x+7x+56)/16 = x
16x – 9x = 56
7x = 56 => x=8cm

##You can also solve this problem using the **convenient number technique**. The convenient number is 8. One-eighth of 8 is 1. Half of the remaining is 3.5.
8-4.5=3.5
Therefore, the answer is 8

4. In a certain shop, 9 oranges cost as much as 5 apples; 5 apples cost as much as 3 mangoes; 4 mangoes cost as much as 9 lemons. If 3 lemons cost 48 paisa, the price of an orange is
Solution:
9 oranges = 5 apples = 3 mangoes
4 mangoes = 9 lemons = 144 paisa
Therefore, 1 mango costs 36 paisa (144/4)
1 mango = 3 oranges
1 orange costs 36/3 = 12 paisa

5. In an examination, 35% of the candidates failed in one subject and 42% failed in another subject, while 15% failed in both the subjects. If 2500 candidates appeared at the examination, how many candidates passed in either subjects but not in both.
Solution:

$$n(A \cup B) = n(A) + n(B) - n(A \cap B)$$

Candidates who failed at least in one subject = 35% + 42% − 15%
= 62%
Therefore, candidates who failed in only one subject = 62% −15%
= 47%
Candidates who failed in only one subject = Candidates who passed in only one subject
Hence 47% of 2500 = 1175

6. In a group of buffaloes and ducks, the number of legs is 24 more than twice the number of heads. What is the number of buffaloes in the group?
Solution:
Let the number of buffaloes be x and ducks be y
Therefore number of legs = 4x + 2y
And number of heads = x + y
Given that 4x + 2y = 2(x + y) + 24
It implies x = 12

7. A certain type of wooden board is sold only in length of multiples of 25 cm from 2 to 10 metres. A carpenter needs a large quantity of this type of boards in 1.65-metre lengths. For the minimum waste, the lengths to be purchased should be:
Solution:
The wooden board is sold in multiples of 25 cm
In addition, the carpenter needs lengths of 165 cm
Therefore, we have to find the LCM of 25 and 165, which is equal to 825 cm or 8.25 metre.

8. If the sum of the digits of an even number is divisible by 9, then that number is always divisible by
Solution:
The first even number, which is divisible by 9, is 18. Hence, all the even numbers, which are divisible by 9, will be multiple of 18. Therefore 18 will divide all such numbers.

9. In a community of 175 persons, 40 read the Times, 50 read the Hindu and 100 do not read any. How many persons read both the papers?
Solution:
Persons who read either the Times or the Hindu = 175 - 100 = 75
Refer problem 5 above.
Persons who read both the papers = 40+50−75 = 15

10. How many 5 digit numbers are there which are divisible by 11 if the five digits are 3, 4, 5, 6 and 7
Solution:
Refer the theory for the divisibility condition of 11
The number so formed is 53647
Now, 5, 6 and 7 can be rearranged in odd places. It can be done in 3! Ways. Similarly 3 and 4 can be arranged in 2 ways. Therefore, we can form 6x2=12 numbers.

11. A printer numbers the pages of a book starting with 1 and uses 3189 digits in all. How many pages does the book have?
Solution:
From 1 to 9 there are 9 digits
From 10 to 99 there are 90x2=180 digits
From 100 to 999 there are 900x3=2700 digits
Therefore, up to 999, the printer will use 2889 digits

The remaining digits are 3189−2889=300 digits
The next page number is 1000, which contains 4 digits.
Hence the remaining pages are 300/4 = 75 pages.
Therefore, the number of pages of the book is 1074

12. Of the 120 people in the room, 3/5th is women. If 2/3rd of the people are married, then what is the maximum number of women in the room who could be unmarried?

Solution:

Of the 120 people, 72 are women and 48 are men.
Of the 120 people, 80 are married.
If all the men are married then the remaining 32 are married women.
Therefore the maximum number of women who are unmarried will be 72−32=40. Note that to maximise the number of unmarried women you have to minimise the married women.

13. Mohan ate half a pizza on Monday. He ate half of the remaining on Tuesday and so on. He followed this pattern for one week. How much of the pizza would he have eaten during the week?

Solution:

1/2+1/4+1/8+1/16+1/32+1/64+1/128 = 127/128 = 99.22%
##You can also apply sum of GP formula, where the first term is ½, common ratio is ½ and number of terms is 7.

14. HCF of 3240, 3600 and a third number is 36 and their LCM is 2^4x3^5x5^2x7^2. Find the third number.

Solution:

- **LCM is the product of all prime factors in their highest powers.**
- **HCF is the product of all the common factors in their lowest powers.**

3240 = 2^3x3^4x5
3600 = 2^4x3^2x5^2
If you look at the LCM, the power of 2 is 4. In 3600, we have the power of 2 as 4. However, we do not have the power of 3 as 5. Hence, the third number should contain that. Similarly, the third number should contain the power of 7 as 2.
In addition, HCF of the three numbers is 36 = 2^2x3^2
Therefore, the third number should also contain 2^2.
Hence, the third number is 2^2x3^5x7^2

15. Students of a class are preparing for a drill. They are made to stand in a row. If 4 students were extra in a row, then there would be 2 rows less. However, there would be 4 more rows if 4 students were less in a row. The numbers of students in the class is:

Solution:

Let the number of rows be r and number of columns be c

Therefore $(r-2)(c+4) = rc$

And, $(r+4)(c-4) = rc$

Hence, $rc+4r-2c-8 = rc-4r+4c-16 = rc$

$4r-2c = 8$ and

$4c-4r = 16$

Solving the above two equations we get, $c = 12$ and $r = 8$

Therefore, the total number of students is 96

16. Which of the following is a square number?

a) 1046535 b) 1046534 c) 1046526 d) 1046529

Solution:

PROBLEM SOLVING TECHNIQUE

Any square number is either divisible by 4 or leaves a remainder 1 when divided by 4. The correct answer is d.

EXERCISE

1. The sum of two numbers is 462 and their highest common factor is 22. What is the maximum number of pairs that satisfy these conditions?

2. If Dennis is $1/3^{rd}$ the age of his father Keith now, and was $1/4^{th}$ the age of his father 5 years ago, then how old will his father Keith be 5 years from now?

3. A man has 1044 candles. After burning, he can make a new candle from 9 stubs left behind. Find the maximum number of candles that can be made.

4. A sum of Rs.36.90 is made up of 180 coins, which are either 10 paisa, or 25-paisa coins. Determine the number of each type of coins.

5. $6500 was divided equally among a certain number of persons. Had there been 15 more persons, each would have got $30 less. Find the original number of persons.

6. One-fourth of a herd of camels was seen in the forest. Twice the square root of the herd had gone to mountains and the remaining 15 camel were seen on the bank of the river. Find the total number of camels.

7. The numbers from 1 to 29 are written side by side as follows:
   1234567891011121314............272829
   If the number is divided by 9 what will be the remainder.

8. The charges for a five-day trip by a tourist bus for one full ticket and a half ticket are Rs.1,440 inclusive of boarding charges which are same for a full ticket and a half ticket. The charges for the same trip for 2 full tickets and one half-ticket inclusive of boarding charges are Rs.2, 220. The fare for a half-ticket is 75% of the full ticket. Find the fare and the boarding charges separately of a full ticket.

9. The number of positive integers not greater than 100, which are not divisible by 2, 3 or 5 is

10. Suppose one wishes to find positive integers (x, y) such that $(x+y)/xy$ is also a positive integer. Identify the correct alternative.

1) This is never possible
2) This is possible and the pair (x, y) satisfying the stated condition is unique.
3) This is possible and there exist more than one but a finite number of ways of choosing the pair
4) This is possible and the pair can be chosen in infinite ways.

11. Let x, y and z be positive integers satisfying $x<y<z$ and $x+y+z = k$. What is the smallest value for k that does not determine x, y and z uniquely?

12. Given odd positive integers x, y and z, which of the following is not necessarily true?

1) $x^2y^2z^2$ is odd
2) $3(x^2+y^2)z^2$ is even
3) $5x+y+z^4$ is odd
4) $z^2(x^4+y^4)/2$ is even

13. How many zeroes will be there at the end when you multiply all the integers from 1 to 100?

14. The rightmost digit in the decimal representation of 2^{51} is.

15. If in $(BE)^2=MPB$, the letters stand for distinct digits, then M equals

SOLUTIONS

Sol: 1

PROBLEM SOLVING TECHNIQUE:

As the HCF of the two numbers is 22, the numbers will be in the form of 22x and 22y where x and y are positive integers and are prime to each other. HCF will be a factor of the two numbers.

Now, $22x + 22y = 462$; => $x + y = 21$

The set of pairs satisfying the conditions are (1, 20), (2, 19), (4, 17), (5, 16), (8, 13), (10, 11)

Sol: 2

Let the age of Dennis be x, therefore Keith will be 3x

5 years ago Dennis = x−5, and Keith = 3x−5

Given that $4(x-5) = 3x-5$; => $x = 15$

Keith's present age is 45; after 5 years, his age will be 50

Sol: 3

From 1044 stubs, he can make 116 candles. From 116 stubs, he can again make 12 candles and there will be 8 stubs remaining. When 12 candles are burnt, there will be 12 stubs left. Now totally he got 20 stubs. From this, he can again make 2 more candles. Therefore, altogether he can make 130 candles.

Sol: 4

Let the number of each type of coins be x and y

Therefore, $x + y = 180$ and

$10x + 25Y = 3690$

$10x + 10y = 1800$ (multiplying the first equation with 10)

Solving these simultaneous equations, we get

$15y = 1890$; => $y = 126$ and therefore $x = 54$

Sol: 5

Let the number of persons be x

Originally, each must have got 6500/x

If there are 15 more persons then each will get only 6500/(x+15). The difference is given as 30.

$(6500/x) - 6500/(x+15) = 30$

$6500[(x+15)-x] = 30(x+15)x$
$6500x15 = 30x^2+450x$, or
$3250 = x^2+15x$
Solving the quadratic equation we get $x = 50$

Sol: 6
Let the number of camels be x. Therefore, x/4 camels were in the forest.
$2\sqrt{x}$ camels had gone to mountains. The remaining 15 were seen on the bank of the river.
Hence $(x/4)+2\sqrt{x}+15 = x$; => $2\sqrt{x} = x-(x/4)-15 = (3x/4)-15$
Squaring both sides we get $4x = 9x^2/16+225-(90x)/4$;
$9x^2-424x+3600=0$
$9x^2-324x-100x+3600=0$ or $(9x-100)(x-36)=0$
Therefore x =36
You can use the **convenient number** technique.
Number of camels is a square number and is divisible by 4. In addition, it is greater than 15. 25 are not divisible by 4. The next number is 36. 9 camels in the forest, 12 in mountains, and the remaining 15 are on the bank of the river.

Sol: 7
PROBLEM SOLVING TECHNIQUE
If the sum of the digits is divisible by 9 then the number is divisible by 9.
Sum from 1 to 9 = 45
Sum from 10 to 19 = (sum from 1to 9 plus 10x1) = 55
Sum from 20 to 29 = (sum from 1 to 9 plus 10x2) = 65
Therefore sum of the digits = 165; if this number is divided by 9 the remainder is 3
You can also think in another way. This sequence contains three times the numbers from 1 to 9. The sum from 1 to 9 is divisible by 9. The remaining numbers are 10 ones and 10 twos. The sum is 30. When you divide it by 9 the remainder is 3.

Sol: 8
It is given that the fare for a half-ticket = 75% of a full-ticket
Therefore, the charges for one and a half ticket = 1.75 x fare+2xboarding charges. Similarly, for two and a half ticket = 2.75 x fare+3xboarding charges.
1440 = 1.75F + 2B
2220 = 2.75F + 3B

Subtracting we get 1F + 1B = 780
Adding we get 4.5F + 5B = 3660
We know 5F + 5B = 3900
From the above two equations we get .5F = 240
Therefore F = 480 and B = 300

Sol: 9
From 1 to 100 there are 50 even numbers, which are divisible by 2. Therefore, 50 odd numbers are left out.
From 1 to 100 there are 33 numbers, which are divisible by 3. In this there are 16 even numbers, which are divisible, by 2 and 17 odd numbers are left out.
That is out of the 50 odd numbers 17 are divisible by 3 and the remaining numbers are 33.
There are 10 odd numbers, which are divisible by 5. Out of this, 15, 45 and 75 are also divisible by 3. Hence, there are only 7 odd numbers, which are divisible by 5. Therefore, we have 33-7 = 26 numbers, which are not divisible by 2, 3 or 5.
Students who find difficult to comprehend the above can simply write down the numbers, which may not take much time.
There are 25 prime numbers from 1 to 100. If you exclude 2, 3, and 5 there are 22 prime numbers, which are not divisible by 2, 3 or 5. In addition 1, 49, 77, and 91 are also not divisible by 2, 3, or 5. Therefore, there are 26 numbers.

Sol: 10
When x = 1 and y = 1 we get a positive integer value 2
When x = 2 and y = 2 we get a positive integer value 1
Therefore, the first 2 options are wrong. For any other value of x and y we always get fractions. Hence, option 3 is correct.

Sol: 11
The sum of the first 3 integers (1+2+3) = 6. This is a unique value.
If k = 7 then you can change the value only in z. That will also be a unique value. Actually we have to execute changes in two of x, y and z.
That is possible only if k = 8.
1+3+4 = 8 and 1+2+5 = 8

Sol: 12
PROBLEM SOLVING TECHNIQUE:
When an even number is divided by 2, the result is either even or odd. For example 2/2 = 1; 4/2 = 2; 6/2 = 3 and so on.

When $x = 1$ $y = 3$ and $z = 5$, the fourth option gives an odd value. Hence, fourth option is not necessarily true.

Sol: 13

PROBLEM SOLVING TECHNIQUE:

2x5 = 10. Therefore, we have to find how many 2's and 5's are there in the product. Definitely, the number of 2's is more than the number of 5's. Hence, we have to find only for 5. There is an easy way of finding that. The method is given below:

100/5 = 20; 20/5 = 4; 4/5 is a fraction.

Add all the quotients, which are integers. 20+4 = 24

(You have to divide the resultant quotients continuously until you get fraction. You do not need to consider the remainders.)

To find how many 2's are there in 100!

100/2 = 50; 50/2 = 25; 25/2 = 12; 12/2 = 6; 6/2 = 3; 3/2 = 1.

$\therefore$ 50+25+12+6+3+1 = 97.

Sol: 14

PROBLEM SOLVING TECHNIQUE:

We have to find the unit digit. The unit digit of

2^1=**2**; 2^2=**4**; 2^3=**8**; 2^4=**6**; 2^5=**2**; 2^6=**4**; 2^7=**8**; 2^8=**6** and so on

If you see, the unit digits of the powers of 2 you will find a pattern. The unit digits are repeated for every 4 powers. (2, 4, 8, 6)

$2^{51} = 2^{48} x 2^3$

Divide 51 by 4. The remainder is 3. $\therefore$ $2^3 = 8$

Hence, the unit digit of 2^{51} is 8.

Sol: 15

PROBLEM SOLVING TECHNIQUE:

For this type of question, **if you know all the square numbers from 1 to 30 it will be easier and that will save a lot of time.**

The first 3-digit number by squaring a 2-digit number is 100 and the last number is 961, which is 31 squared. Therefore, BE has to be one of those numbers from 10 to 31. Hence, B takes the value from 1 to 3. If B is 1 then the unit digit of the three-digit number is 1. There are only two such numbers, 11 and 19. $11^2 = 121$; $19^2 = 361$. The number 11 is not possible because the letters represent distinct digits. Therefore, B=1; E=9; M=3; P=6.

Hence, the rightmost digit will be 8.

3. PERCENTAGE/PROFIT AND LOSS/SI AND CI

PERCENTAGE

x% means x parts out of 100 parts

Therefore x% = x/100
10% = 10/100 = 1/10 = .1
15% = 15/100 = .15
1% = 1/100 = .01
1.5%= 1.5/100=.015
110% = 110/100 = 1.1
90% = 90/100 = .9
10% of x = (10/100)x = .1x

If the price of a commodity is increased by 10%, to find the final price, simply multiply the price with 1.1

If the price of a commodity is decreased by 10%, to find the final price, simply multiply the price with .9

If the price of the commodity is successively increased by 10% and 20%, to find the final price, simply multiply the price with 1.1x1.2

If the price of the commodity is first increased by 10% and then decreased by 20%, to find the final price, simply multiply the price with 1.1x.8. Note that the result will be the same even if increase and decrease are interchanged.

If the price of the commodity increases by r%, then the percentage of reduction in consumption, expenditure being constant, is given by a formula
[r/(r+100)]x100

If the price of the commodity decreases by r%, then the percentage of increase in consumption, expenditure being constant, is given by a formula
[r/(100-r)]x100

If the length of a rectangle is increased by x% and the breadth is increased by y% then the percentage increase in area is given by a formula x+y+(xy/100). Note that for a decrease a negative variable is used. In the above example if the breadth is decreased by y% then the formula is $x-y+(-xy/100)$. For all two dimensional geometric figures this formula can be used.

PROFIT AND LOSS

Profit = selling price - cost price
Loss = cost price - selling price
Profit% = (profit/cost price)x100
Loss% = (loss/cost price)x100
If a person sells two similar products one at a gain of x% and another at a loss of x% then he always incurs a loss. The loss percentage is given by a formula $x^2/100$. Note that we have used the same formula $x+y+xy/100$, where both the variables are x and one is negative.

SIMPLE INTEREST

If the interest on the money borrowed or lent is reckoned uniformly throughout the period then it is called simple interest. Remember that the interest is never added to the principal for the purpose of calculation of interest. Interest is reckoned only on the principal. Simple interest is given by the formula,
SI = Pnr/100, where P is the principal or sum, n is the time period in years and r is the rate of interest per annum.
Amount = Principal + SI

COMPOUND INTEREST

Interest is generally calculated for a certain unit time. If the interest is added to the principal for reckoning interest for the next unit time then it is called compound interest. The unit time may be quarterly, half-yearly or yearly. Compound interest is given by a formula:

$$CI = P[1+\frac{r}{100}]^n - P$$

$$Amount = P[1+\frac{r}{100}]^n$$

Where P is the principal, r is the rate of interest per annum and n is the number of times.
If the interest is reckoned quarterly then r is to be divided by 4 and n is to be multiplied by 4
If the interest is reckoned half-yearly then r is to be divided by 2 and n is to be multiplied by 2

SOLVED EXAMPLES

1. When the price of a radio was reduced by 20%, its sale increased by 80%. What was the net effect on the sale?
Solution:
Let the price of the radio be x and the quantity of its sale be y
Then, (.8x)(1.8y) will be the current total sales, which is equal to 1.44xy
Therefore, there is an increase of .44xy sales
This is equal to 44% increase.
You can apply the formula (x+y+xy/100)
-20+80-16=44

2. In a certain store, the profit is 320% of the cost. If the cost increases by 25% but the selling price remains constant, approximately what percentage of the selling price is the profit?
Solution:
Let the cost price be 100. Therefore selling price = 100+320=420.
Now the cost price is increased to 125 and the SP =420
Profit = 420−125 = 295
To find x% of 420 = 295
(x/100)420 = 295
x = (295x100)/420 = 70% approximately.

3. If two numbers are respectively 20% and 50% of a third number, what is the percentage of the first number to the second?
Solution:
Let the third number be 100
Therefore, the first and second numbers are 20 and 50 respectively.
We have to find the percentage of the first to the second.
We are comparing the first to the second.
The ratio of first to the second is 20/50.
To find the percentage, multiply with 100
Therefore, the required answer is 40%
In percentage problems, the **convenient number** is 100

4. The prices of two houses A and B were 4:5 last year. This year, the price of A is increased by 25% and that of B by

Rs.50000. If their prices are now in the ratio 9:10 then what is the price of the house A last year.

Solution:

Let the price of A and B be 4x and 5x respectively.

This year the price of A and B is 1.25(4x) = 5x and 5x+50000

They are in the ratio 9:10

9 -> 5x

10 -> 5x+50000

If you cross multiply and equate you will get x = 90000

Therefore, the price of A last year was 360000

5. If a certain sum of money becomes double at simple interest in 12 years, what would be the rate of interest per annum?

Solution:

##Divide 100 by the number of years then you get the answer for this type of problems. The required answer is 8.33%

6. A shopkeeper sells two radios at Rs.1540 each. On one, he gains 12% and on the other, he loses 12%. What was the net result of the sale of both the radios?

Solution:

$$Refer the theory. The formula given there is x+y+xy/100.

12–12–(12x12)/100 = –1.44%

Negative indicates loss.

##The shopkeeper will always incur loss.

To find the amount:

If the CP is 100 and the loss, is 1.44 then SP will be 98.56

Now 98.56 -> 3080

1.44 -> ?

If you cross multiply, you get the answer.

Loss amount = (3080x1.44)/98.56 or (3080x144)/9856 = 45

Alternative method:

Find the cost price of each then the difference of total SP and total CP will be the answer. Try yourself. It will take more time.

7. Two equal sums were borrowed at 8% simple interest per annum for 2 years and 3 years respectively. The difference in the interests was Rs.56. The sums borrowed were:

Solution:

The difference of Rs.56 is the interest for one year at 8% simple interest.

Therefore, 8% -> 56

100% -> x

If you cross multiply you get the sum and x = 700

8. By selling 12 marbles for a rupee, a shopkeeper loses 20%. In order to gain 20% in the transaction, he should sell the marbles at the rate of how many marbles for a rupee?

Solution:

SP of 12 marbles = 100paise; loss = 20%
CP of 12 marbles = 100(100/80) = 125paise
In order to gain 20%, SP = 125(120/100) = 150paise
So, the selling price of 12 marbles = 150paise
Therefore, for 100paise 12(100/150) = 8 marbles should be sold.
##If the shopkeeper sold 10 marbles for a rupee, she would not have incurred 20% loss (cost price of 10 marbles=selling price of 10 marbles). If she sells 8 marbles for a rupee she will gain 20%.

9. At what percentage above the cost price must an article be marked to gain 33% after allowing a customer a discount of 5%?

Solution:

If the cost price is 100, to gain 33% it should be sold at 133.
This selling price of 133 is after allowing a discount of 5%
That is if the marked price is 100 then the selling price is 95.
Therefore, 95 -> 133
100 -> x
If you cross multiply you get x = 140.
Hence, the marked price is 40% above the cost price.
The **convenient number** is 100. Cost price=100, selling price=133. The marked price is greater than 133. 5% of an amount greater than 133 is greater than 5% of 133 (6.65). If you take an approximate value of 7, you get 140 which is the answer.

10. Successive discounts of 10%, 12% and 15% amount to a single discount of:

Solution:

If the marked price is 100 then after the first discount of 10% it will become 90, after the second discount of 12% it will become 79.20 and after the third discount of 15%, it will become 67.32. Therefore, single discount equals to
100 – 67.32 = 32.68%
##Note that for a discount of 12% multiply with .88 and for a discount of 15% multiply with .85

EXERCISE

1. A piece-of-cloth costs Rs.35. If the length of the piece had been 4m longer and each metre, costs Rs.1 less, the cost would have remained unchanged. How long is the piece?

2. A horse and a carriage together cost Rs.8000. If by selling the horse at a profit of 10% and the carriage at a loss of 10% a total profit of 2.5% is made, then what is the cost price of the horse?

3. In a co-educational school, there are 15 more girls than boys. If the number of girls is increased by 10% and the number of boys is also increased by 16% there would be nine more girls than boys. What is the number of students in the school?

4. A sum of money becomes eight times in 3 years if the rate is compounded annually. In how much time the same amount at the same compound interest rate will become sixteen times?

5. A loss of 19% is converted into a profit of 17% when the selling price is increased by Rs.162. Find the cost price of the article.

6. The digit at unit's place of a two-digit number is increased by 100% and the ten's digit of the same number is increased by 50%.The new number thus formed is 19 more than the original number. What is the original number?

7. A family wanted to reduce the expenditure on milk by 19%. However, the price of milk increased by 8%. In order to achieve the desired result, by what percentage the family would reduce the consumption?

8. A trader professes to sell his goods at a loss of 10% but gives only 800 grams instead of 1 kg. What is his gain percentage?

9. A man invests Rs.5000 for 3 years at 5% p.a. compound interest reckoned yearly. Income tax at the rate of 20% on the interest earned is deducted at the end of each year. Find the amount at the end of third year.

10. Peter got 30% of the maximum marks in an examination and failed by 10 marks. However, Paul who took the same examination got 40% of the total marks and got 15 marks more than the passing marks. What were the passing marks in the examination?

11. Mr. X's salary is increased by 20%. On the increase, the tax rate is 10% higher. The percentage increase in tax liability is:

SOLUTIONS

Sol: 1

Let the length be L and the price per metre be P
Given that LP = 35
(L+4)(P−1) = 35; => LP−L+4P−4 = 35
Therefore, 35−L+4P−4 = 35
=> 4P−L = 4 or L = 4P−4 = 4(P−1)
Hence, 4(P−1)P = 35
$4P^2-4P-35 = 0$
(2P+5)(2P−7) = 0
Therefore, P = 3.50 and L = 10

Sol: 2

Let the cost price of horse be x and the cost price of carriage be y
Therefore, x +y = 8000
Given that, 1.1x +.9y = 8200 (2.5% of 8000 = 200)
Now multiply the first equation with .9
Therefore, .9x+.9Y = 7200
1.1x+.9y = 8200
It implies .2x = 1000; x = 5000

Sol: 3

Let the number of boys be x, then the number of girls will be x+15
Given that, 1.1(x+15)−1.16x = 9
1.1x+16.5−1.16x = 9 or .06x = 7.5
Therefore, x = 125
Total number of students = 125+125+15 = 265

Sol: 4

Let the sum be P. It becomes 8P in 3 years.
$PR^3 = 8P$; where R = 1+ (r/100)
$R^3 = 8$; R = 2
Therefore, r = 100% => every year it doubles.
Hence, it will become 16P in the next year. Therefore, it will take 4 years.

Sol: 5

PROBLEM SOLVING TECHNIQUE:

Let the cost price be 100. If the loss is 19% then the selling price is 81. If it is sold at a profit of 17% then the selling price is 117.
Now the difference between the two selling prices is 36.

If the difference is 36, the cost price is 100.
Therefore, if the difference is 162 then the cost price will be:
36 -> 100
162 -> x
If you cross multiply, x = 450

Sol: 6
Let the number be xy. The new number is 1.5x2y.
Given that 1.5x2y – xy = 19
xy is nothing but 10x+y
Applying the above rule, we get
(15x+2y)–(10x+y) = 19
5x+y = 19
5x = 19––y
The unit digit y can take the value of 4 or 9
Therefore, x = 3 and y = 4
Or x = 2 and y = 9
The numbers can be 34 or 29

Sol: 7
Let the family buy 1000 ml of milk at Rs.10
To cut the expenditure by 19% it would buy only 810 ml of milk for Rs.8.10
However, the price of milk is also increased to Rs.10.80
Rs.10.80 -> 1000 ml
Rs.08.10 -> x
If you cross multiply you get x = 750 ml
That implies the family would have to reduce the consumption by 25%

Sol: 8
Let the cost price be 100 for 1000 grams
The trader sells at 90, 10% loss
However, the trader gives only 800 grams.
The cost price of 800 grams = 80
The selling price of 800 grams = 90
Therefore, gain% = (10/80)100 = 12.5%

sol: 9
Tax is deducted from the income. Therefore, income tax rate on the sum will be equal to 20% of 5% = 1%. Hence, the net effective rate of interest is 4%.
Amount = PR^n
Amount = 5000x1.04x1.04x1.04 = 5788.125

Sol: 10

PROBLEM SOLVING TECHNIQUE

The difference in percentage of marks obtained by Peter and Paul is 10%

The difference in actual marks obtained by them is 25. Therefore,

10% -> 25

100% -> x

If you cross multiply you get x = 250

Therefore, the total marks are 250 and passing marks are 85.

Sol: 11

It cannot be found because the tax rate on the initial salary is not given.

4. RATIO, PROPORTION, AVERAGE & PARTNERSHIP

RATIO AND PROPORTION

Ratio is nothing but a comparison of two quantities. If A's weight is 'a' and B's weight is 'b', then the ratio of A's weight to B is a:b or a/b

When two ratios are equal, they are in proportion.
a/b = c/d =>ad=bc
'a' is called first proportion, 'b' is called the second proportion, 'c' is called the third proportion and 'd' is called the fourth proportion.
a/b=c/d => b/a=d/c

- ⇨ a/c=b/d
- ⇨ (a+b)/b=(c+d)/d
- ⇨ (a−b)/b =(c−d)/d
- ⇨ (a+b)/(a−b) = (c+d)/(c−d)
- ⇨

The compounded ratio of a:b and c:d is ac:bd

If a:b::c:d, then d is called the fourth proportion

If a:b::b:c, then c is called the third proportion and b is called the mean proportion and b = $\sqrt{ac}$

Continued proportion

If $\frac{a}{b}=\frac{c}{d}=\frac{e}{f}=.......$ then $\frac{a+c+e+....}{b+d+f+....}=\frac{a}{b}=\frac{c}{d}=\frac{e}{f}=.....$

VARIATION

If x is directly proportional to y then x = ky, where k is a constant.
If x is inversely proportional to y then x = k/y, where k is a constant.

PARTNERSHIP

When two or more persons jointly run a business, they are called partners and the entity is called partnership firm.

When the partners invest their capital for a fixed period, say one year, the profit of that period is divided among the partners in the ratio of their capital invested.
When the partners invest their capital for different times then the profit is divided among the partners in the compounded ratio of time and capital.

If the capital ratio is x:y:z and the time ratio is a:b:c then the profit ratio is
xa:yb:zc

AVERAGE

Average = sum of observations /number of observations
Arithmetic mean of two numbers a and b = (a + b)/2
AM of three numbers a, b and c = (a + b + c)/3
Geometric mean of two numbers a and b = $\sqrt{ab}$
GM of three numbers a, b and c = $\sqrt[3]{abc}$
Harmonic mean of two numbers a and b = 2ab/(a + b)
HM of three numbers a, b and c =3/[1/a+1/b+1/c]
AM≥GM≥HM
GM^2 = AM X HM

SOLVED EXAMPLES

1. The total emoluments of A and B are equal. However, 'A' gets 65% of his basic salary as allowances and B gets 80% of his basic salary as allowances. What is the ratio of the basic salaries of A and B?

Solution:

PROBLEM SOLVING TECHNIQUE

If A's basic salary is 100 then his emoluments will be 165
If B's basic salary is 100 then his emoluments will be 180
Let the emoluments of A and B be x each.
A's basic salary = 100x/165
B's basic salary = 100x/180
Therefore, A/B = 180/165 = 12/11

2. A contractor employed 25 labourers on a job. He was paid Rs.275 for the work. After retaining 20% of this sum, he distributed the remaining amount amongst the labourers. If the number of men to women labourers was in the ratio 2:3 and their wages in the ratio 54 what wages did a woman labourer get?

Solution:

The amount distributed among the labourers is .8x275 = 220
$$The compounded ratio of 2:3 and 5:4 is 10:12
The total wages earned by men and women are 100 and 120
The number of women labourers is 15
Therefore, each woman got Rs.8

3. An alloy contains copper and zinc in the ratio 5:3 and another alloy contains copper and tin in the ratio 8:5. If equal weight of both the alloys are melted together then the weight of tin in the resulting alloy per kg will be

Solution:

PROBLEM SOLVING TECHNIQUE:

Sum of the first ratio is 8 and the sum of the second is 13. Therefore, multiply the first ratio with 13 and the second with 8 to make the two alloys equal in weight. Now the ratio of the first is 65:39 and the second becomes 64:40. Now, if you mix up you will get the weight of copper, zinc and tin as 129, 39 and 40 respectively. The total weight of the resulting alloy is 208. In this the weight of tin is 40/208 or 5/26.

4. Atul and Babita enter into a business partnership in which Atul contribute Rs.2000 for 9 months and Babita contributes Rs.5000 for 7 months. A profit of Rs.1100 will be divided between Atul and Babita in the ratio of:

Solution:

$$The profit ratio will be the compounded ratio of capital and time ratio.

Therefore, the profit ratio is 18000:35000 or 18:35

5. A tin of oil was four-fifths full. When six bottles of oil were taken out and four bottles of oil were poured into it, it was three-fourths full. How many bottles of oil were contained by the tin?

Solution:

The net effect is 2 bottles were taken out. The tin is reduced from 4/5 to ¾.

4/5 – 3/4 = 1/20

1/20 -> 2 bottles

1-> x

If you cross multiply you will get x = 40 bottles.

6. Ram spends Rs.3620 for buying pants at the rate of Rs.480 each and shirts at the rate of Rs.130 each. What will be the ratio of pants to shirts when maximum number of pants is to be bought?

Solution:

PROBLEM SOLVING TECHNIQUE

> **In this problem we are facing two variables and only one equation. We cannot solve this using the normal simultaneous equations' rules. This is called special equation. You have to apply the following method:**

Let the number of pants be x and shirts be y

Therefore, 480x+130y = 3620 or 48x+13y = 362

48x = 362−13y or x = (362−13y)/48 = (336+26−13y)/48

Since we have to maximise x give minimum value for y, which is equal to 2.

When y=2, x=7. Therefore, the ratio is 7:2

7. If Rs.1066 are divided among A, B, C and D such that A:B=3:4, B:C=5:6 and C:D=7:5, who will get the maximum?

Solution:

If you combine the first two ratios, A:B:C=15:20:24; C:D=7:5

If you combine the above ratios, A:B:C:D=105:140:168:120. Therefore, C gets the maximum.
In the first ratio the value of B is 4 and in the second ratio the value is 5. Take LCM of 4 and 5, which is equal to 20. Multiply the first ratio with 5 and the second ratio with 4. Then the value of B will be equal in both the ratios.

8. A man purchased 4 pairs of black socks and some pairs of brown socks. The price of a black pair is double that of a brown pair. While preparing the bill the clerk interchanged the number of black and brown pairs by mistake. Consequently, the bill was increased by 50%. The ratio of the number of black and brown pairs of sock in the original order was:
Solution:
Let the price of a pair of brown socks be 1, then black socks will be 2. Also let the number of pairs of brown socks is n.
Therefore, actual bill is 8+n and the miscalculated bill is 2n+4
Now, 8+n:2n+4::2:3 => 24+3n = 4n+8 or n = 16
The required ratio is 1:4

9. If P varies as QR and the three corresponding values of P, Q and R are 6, 9 and 10 respectively, then the value of P when Q=5 and R=3 is:
Solution:
If P varies as QR means P = k(QR), where k is constant.
Therefore, 6 = kx9x10 => k = 6/90 = 1/15
Now, P = $\frac{1}{15}$x5x3 = 1

10. The difference between a two digit number and the number obtained by interchanging the digits is 36. What is the difference between the sum and the difference of the digits of the number if the ratio between the digits of the number is 1:2?
Solution:
In the above problem if the number 36 is divided by 9 we get 4. Therefore, the difference between the two digits will be 4. The possible numbers are 15, 26, 37, 48 and 59. Out of these numbers only 48 has the digits in the ratio 1:2. Hence the required answer is 12−4 = 8.

PROBLEM SOLVING TECHNIQUE

> **If the difference between a number and the number obtained when the digits are interchanged, is divided by 9 the quotient will be the difference of the digits of that number.**

11. In three numbers, the first is twice the second and thrice the third. If the average of these numbers is 44, then the first number is:

Solution:
Let the first number be 6n then the second and the third numbers are 3n and 2n respectively.
Average of these numbers is 11n/3 = 44
n = 44x3/11 = 12
Therefore the first number is 72.

12. The batting average of 40 innings of a cricket player is 50 runs. His highest score exceeds his lowest score by 172 runs. If these two innings are excluded the average of the remaining 38 innings is 48. His highest score was:
Solution:
The total runs in 40 innings = 2000
The total runs in 38 innings = 1824
Therefore, the highest and lowest runs together is 176
Let n be the highest score. Therefore, the lowest score will be 176−n
n−176+n = 172
=> n = 174.

13. Nine persons went to a hotel for taking their meals. Eight of them spent Rs.12 each over their meals and the ninth spent Rs.8 more than the average expending of all the nine. Total money spent by them was:
Solution:
8 persons spent a total of Rs. 96
Let the ninth person spent Rs. n
The average of 9 persons = (96+n)/9
Therefore, n−8 = (96+n)/9; => n = 21
The total money spent by them is Rs.117

14. A batsman makes a score of 98 runs in the 19th innings and thus increases his average by 4. What is his average after 19th innings?

Solution:

Let his average be n in 18 innings

Therefore, 18n+98 will be his total runs in 19 innings

The average for 19 innings = (18n+98)/19 = n+4; => n = 22

Hence, his average for 19 innings is 26

ALTERNATIVE METHOD

(Score in n^{th} innings) – (n−1) x increase in average = average in n innings.

98−18x4 = 26

15. If $\frac{y}{x-z} = \frac{y+x}{z} = \frac{x}{y}$ then find x:y:z

Solution:

$$Applying **continued proportion** property, $\frac{2y+2x}{x+y} = \frac{x}{y}$; => $\frac{x}{y} = \frac{2}{1}$

$\frac{y+x}{z} = \frac{x}{y}$

=> $\frac{1+2}{z} = \frac{2}{1}$; => z = $\frac{3}{2}$

Therefore, the required ratio is 4:2:3

EXERCISE

1) Several litres of acid were drawn off a 54 litre vessel full of acid and an equal amount of water added. Again the same volume of the mixture was drawn off and replaced by water. As a result the vessel contained 24 litres of pure acid. How much of the acid was drawn off initially?

2) Rs.770 has been divided among A, B and C such that A received $2/9^{th}$ of what B and C together receive. Then A's share is:

3) One variety of tea worth Rs.126 per kg and another variety of tea worth Rs135 per kg are mixed with a third variety in the ratio 1:1:2. If the mixture is worth Rs.153 per kg then the price of the third variety per kg:

4) In a mixture of 45 litres, the ratio of milk and water is 3:2. How much water must be added to make the ratio 9:11?

5) The monthly incomes of two persons are in the ratio of 4:5 and their monthly expenditures are in the ratio 7:9. If each saves Rs.50 a month then what are their monthly incomes?

6) A bag contains Rs.216 in the form of one rupee, 50 paisa and 25 paisa coins in the ratio of 2:3:4. The number of 50 paisa coins is:

7) The ratio of rate of flow of water in pipes varies inversely as the square of the radius of the pipes. What is the ratio of the rates of flow in two pipes of diameters 2 cm and 4 cm?

8) A started a business with Rs.4500 and another person B joined after some period with Rs.3000. Determine this period before B joined the business if the profit at the end of the year is divided in the ratio 2:1

9) Three containers A, B and C are having mixtures of milk and water in the ratio 1:5, 3:5 and 5:7 respectively. If the capacities of the containers are in the ratio 5:4:5 then find the ratio of the milk to the water if the mixtures of all the three containers are mixed together.

10) From a barrel containing 500 ml of alcohol, 3 cups of alcohol are poured into a barrel containing 500 ml of water. After mixing the contents well, 3 cups of the mixture are poured into the barrel of alcohol. The percentage of water in the barrel of alcohol and the percentage of alcohol in the barrel of water are then compared. Which one of the following is true?

a) The former is greater than the latter
b) The two are equal
c) The latter is greater than the former
d) Cannot be determined.

11) The average weight of 45 students in a class is 52 kg. 5 students whose average is 48 kg leave the class and 5 students whose average is 54 kg join the class. What is the new average of the class?

12) A car owner buys petrol at Rs.7.50, Rs.8.00 and Rs.8.50 per litre for three successive years. What approximately is the average cost per litre of petrol if he spends Rs.4000 each year?

13) The average of three numbers is 135. The largest number is 180 and the difference between the others is 25. The smallest number is:

14) The average of 11 numbers is 10.9. If the average of the first six numbers is 10.5 and that of the last six numbers is 11.4, then the middle number is:

SOLUTIONS

Sol: 1

@@You have a formula for this problem.

$\left(\frac{a-b}{a}\right)^n = \frac{liquidleft}{initialvolume}$; a is the original quantity of the liquid, b is the quantity of liquid drawn off each time and n is the number of operations.

Here a = 54 and b is not given, which we have to find; quantity of acid left = 24. Substitute the above values in the formula.

$$\left(\frac{54-b}{54}\right)^2 = \frac{24}{54}$$

$(54-b)^2 = (24x54x54)/ 54 = 24x54 = 2^4x3^4$

54−b = 36

=> 18

Sol: 2

Let B and C together get x, then A gets 2x/9

x+ 2x/9 = 770 or 11x/9 = 770 => x = (770x9)/11 = 630

Therefore, A gets 140

Sol: 3

Let the price of the third variety be p

1x126 + 1x135 + 2xp = 4x153 = 612

2p = 612-261 = 351

p = 175.50

Sol: 4

In 45 litres, milk = 27 litres and water = 18 litres

Let n litres of water are added to the mixture. Therefore, in the new mixture there will be 18+n litres of water.

If the number 27 is divided by 3 you get 9. Therefore, if 18+n is divided by 3 you should get 11. Hence, 18+n = 33; => n = 15

Sol: 5

PROBLEM SOLVING TECHNIQUE

If you multiply the first ratio by 2 you get 8:10. Now the difference in the ratios is a constant, which is 1. Therefore,

1 → 50

8 → 400

10→ 500

Alternative method

Let the incomes be 4x and 5x and the expenditures be 7y and 9y

4x−7y=50   or 20x−35y=250

5x−9y=50   or 20x−36y=200

=>y=50 and x=100

Therefore, their incomes are 400 and 500

Sol: 6

Let the number of one rupee coins be 2, 50 paisa coins be 3 and 25 paisa coins be 4. Therefore, 2x100+3x50+4x25 = 450 paisa

450 ➔ 9 coins

21600 ➔ n

n = (9x21600)/450 = 432 coins

Therefore, number of 50 paisa coins = (432x3)/9 = 144

Sol: 7

PROBLEM SOLVING TECHNIQUE

The ratio of squares of the diameters is 4:16 or 1:4

Since the rate of flow is inversely proportional to the square of the radius of the pipes, take the inverse ratio that is 4:1.

Sol: 8

PROBLEM SOLVING TECHNIQUE

A's capital is for 12 months. Therefore, 12x4500 = 54000

Half of 54000 = 27000

27000/3000 = 9; => B's capital is for 9 months. Hence, he joined the business after 3 months.

Alternate method:

Let B joined after n months.

Therefore, 12x4500:(12-n)3000::2:1

54000:36000-3000n::2:1

54000 = 72000-6000n

=>n = 3

Sol: 9

	milk	Water
A	1/6	5/6
B	3/8	5/8
C	5/12	7/12

The table shows the quantity of milk and water if one litre from each container is considered. If 5 litres from A, 4 litres from B and 5 litres from C is considered then the quantity of milk and water will be as follows:

	milk	Water
A	5/6	25/6
B	12/8	20/8
C	25/12	35/12

Now if you add milk and water separately you will get the ratio.
Milk = 106/24 and water = 230/24
Therefore, the required ratio is 53:115

Sol: 10

PROBLEM SOLVING TECHNIQUE
As you do not know the capacity of the cups, take some arbitrary value and check with the options.

The two are always equal.
Suppose you remove 100 ml from A and pour this into B. Now B will contain water and alcohol in the ratio 5:1. If you remove 100 ml from B then that 100 ml will contain water and alcohol in the ratio 5:1. That is, the quantity of water is 100x5/6 or 500/6 and alcohol is 100/6. Therefore the quantity of water, which goes to A, is 500/6. Alcohol in B will be 100−(100/6) = 500/6. As the quantity of water in A and the quantity of alcohol in B is the same, the percentage will be the same. Students are advised to check with different quantity.

Sol: 11
The difference in average of 5 students leaving and joining is 6. The net increase in total weight is 5x6 = 30. Therefore, the net increase in average is 30/45 = 2/3.
Hence, the new average is 52.67 kg.

Sol: 12

$$\text{Average} = \frac{12000}{4000\left(\frac{2}{15}+\frac{1}{8}+\frac{2}{17}\right)} = \frac{6120}{767} = 7.98$$

This is nothing but harmonic mean of 7.50, 8 and 8.50.

Sol: 13

The sum of all the 3 numbers is 405.
The total of 2 numbers less the largest number is 405−180 = 225
Let the two numbers be x and x−25
Therefore, x+x−25 = 225; => x = 125
The smallest number is 100

Sol: 14

The sum of 11 numbers is 119.9
The sum of first 6 numbers is 63
The sum of last 6 numbers is 68.4
Therefore, the middle number is 63+68.4−119.9 = 11.5

5. WORK AND TIME

If A can do a work in x days then in one day he can do 1/x part of that work.

If B can do a work in y days then in one day he can do 1/y part of that work.

If A and B together do that work, in one day they can complete $1/x + 1/y$ part of that work or $(x+y)/xy$. Therefore A and B together will complete the whole work in $xy/(x+y)$ days

If the efficiency ratio of A and B is x:y then the ratio of time taken to complete a work by A and B will be y:x or simply efficiency ratio is the inverse of time ratio and vice versa.

If pipe A fills a tank in x hours and pipe B drains the tank in y hours then, when the pipes are opened simultaneously, the tank will be filled in $xy/(y-x)$ hours. Note that $y>x$, otherwise the tank will never be filled.

SOLVED EXAMPLES

1. 5 persons working eight hours daily can complete a wall in 10 days. When they have worked for 5 and a half day, 5 more persons are brought to work. The wall can now be completed in:

Solution:

Suppose the work is divided into 10 equal parts, in 5 and a half day, 5.5 parts of the work will be completed. The remaining work is 4.5 parts.

5 persons 4.5 days 4.5 parts
10 n days 4.5 parts

Since the number of persons is increased, to complete the same work, they will take less number of days. They are in inverse proportions. So, do direct multiplication and equate.

$5x4.5 = 10xn$
$n = 2.25$ days.

2. A tap can fill a cistern in 8 hours and another can empty it in 16 hours. If both the taps are opened simultaneously, the time to fill the cistern will be:

Solution:

Pipe A can fill 1/8 parts of the cistern in 1 hour (whole part in 8 hours)
Pipe B can empty 1/16 parts of the cistern in 1 hour
Together $1/8 - 1/16 = 1/16$ parts in 1 hour (negative stands for emptying)
Therefore, the cistern will be filled in 16 hours.

3. A and B together can do a piece of work in 12 days. B and C together can do the same work in 16 days. After A has worked for 5 days and B worked for 7 days, C completed the work in 13 days. In how many days will C alone be able to do the work?

Solution:

A and B together can do in one day 1/12 parts.
B and C together can do in one day 1/16 parts.
Combine B's 5 days work with A and 2 days work with C.
Now, $5(1/12)+2(1/16)+11/C = 1$
$11/C = 1-(26/48) = 22/48$
$C = 24$ days

4. Pipes A and B can fill a tank in 5 and 6 hours respectively. Pipe C can empty it in 12 hours. The tank is half full. All the three pipes are in operation simultaneously. After how much time will the tank be full?

Solution:

A, B and C together in 1 hour can fill (1/5)+(1/6)−(1/12) = 17/60 part of the tank

Time = remaining part ÷ the rate at which it is filled

Therefore, (1/2)÷(17/60) = 30/17 hours.

5. In climbing a 21 metre long pole, a monkey climbs 6 metre in the first minute and slips 3 metre in the next minute. What time the monkey would take to reach the top of the pole?

Solution:

In 2 minutes it climbs 3 metre. Therefore, in 10 minutes it can climb 15 metres.

In the next minute it will climb 21 metres.

EXERCISE

1. 24 men working at 8 hours a day can finish a work in 10 days. Working at the rate of 10 hours a day, the number of men required to finish the same work in 6 days is:

2. A certain job was assigned to a group of men to do it in 20 days. However, 12 men did not turn up for the job and the remaining men did the job in 32 days. The original number of men in the group was:

3. A mother and a daughter working together can complete a work in 4 days. However, if the mother works alone she can complete the work in 6 days. Both of them worked for one day and then the mother had to leave. How long will the daughter take to complete the remaining work?

4. If 15 women or 10 men can complete a project in 55 days, in how many days will 5 women and 4 men working together complete the same project?

5. Ramesh is twice as good worker as Sunil and finished a piece of work in 3 hours less than Sunil. In how many hours they together could finish that piece of work?

6. A can do a work in 18 days, B in 9 days and C in 6 days. A and B start working together and after 2 days C joins them. What is the total number of days taken to finish the work?

7. Pipes A and B running together can fill a cistern in 6 minutes. If B takes 5 minutes more than A to fill the cistern then if the pipes are operated separately how much time each will take to fill the cistern?

8. Two men and 7 children complete a certain piece of work in 4 days while 4 men and 4 children complete the same work in only 3 days. The number of days required by 1 man to complete the work is:

9. Rohit, Harsha and Sanjeev are three typists who working simultaneously can type 216 pages in four hours. In one hour Sanjeev can type as many pages more than Harsha as Harsha can type more than Rohit. During a period of five hours Sanjeev can type as many

pages as Rohit can during seven hours. How many pages does each of them type per hour?

10. A contractor undertakes to build a wall in 50 days. He employs 50 people for the same. However after 25 days he finds that only 40% of the work is completed. How many more men needed to complete the work in time?

SOLUTIONS

Sol: 1

24 men   8 hours   10 days   1 work
n men   10 hours   6 days   1 work

PROBLEM SOLVING TECHNIQUE

$$Men and hours are in inverse proportion. Therefore multiply directly.
$$Men and days are also in inverse proportion. Therefore multiply directly.
For direct proportion do cross multiplication.

$$n = 24 \times \frac{8}{10} \times \frac{10}{6} = 32 men$$

Sol: 2

Let the original number of men be n.

n men → 20 days
(n−12) men→ 32 days

$$Men and days are in inverse proportion.

Therefore, $n \times \frac{20}{32} = n - 12$

20n = 32n−384
12n = 384; n = 32

Sol: 3

The rate at which the daughter works = 1/4 − 1/6 = 1/12
In one day both working together can complete 1/4 part of the work.
The remaining part of the work = 3/4.
Therefore, time taken by the daughter = 3/4 ÷ 1/12 = 9 days.

@@ $Time = \frac{work, left}{Rate}$

Sol: 4

PROBLEM SOLVING TECHNIQUE

Given that 10 men = 15 women => 1 man = 15/10 women = 3/2 women.

Therefore, 4 men = 6 women.

5 women + 4 men = 11 women.
Now, 15 women   55 days
11 women   n days

They are in inverse proportion. Therefore, multiply directly.

$15 \times 55 = 11 \times n$
$n = 75$ days.

Sol: 5
PROBLEM SOLVING TECHNIQUE
The efficiency ratio of Ramesh and Sunil is 2:1. Therefore, the time taken by them to do a certain work will be in the ratio 1:2.
If Ramesh takes 3 hours then Sunil will take 6 hours. Working together they can complete $1/3 + 1/6 = 3/6$ or $1/2$ part of the work.
Hence, they together complete the work in 2 hours.

Sol: 6
Let the total number of days be n.
Therefore, $n(1/18 + 1/9) + (n-2)1/6 = 1$
$n(1/6) + n/6 - 1/3 = 1$
$2n/6 = 4/3$
$n = (4/3) \times 3 = 4$

Sol: 7
Let A takes n minutes and B takes n+5 minutes.
$1/n + 1/(n+5) = 1/6 \Rightarrow n = 10$

Sol: 8
PROBLEM SOLVING TECHNIQUE
2 men + 7 children = 4 days or 8 men + 28 children = 1 day
4 men + 4 children = 3 days or 12 men + 12 children = 1 day
That implies 8 men + 28 children = 12 men + 12 children
4 men = 16 children or 1 man = 4 children
Substitute in the second equation
5 men 3 days
1 man n days
$\Rightarrow n = 15$ days

Sol: 9
$S - H = H - R$ or $5S - 5H = 5H - 5R$
Also $5S = 7R$
Therefore, $7R - 5H = 5H - 5R$; $\Rightarrow 12R = 10H$ or $R = 5H/6$
$H - R = H - 5H/6 = H/6$
$S - H = H/6$; $\Rightarrow S = 7H/6$
Together in one hour they can type $7H/6 + H + 5H/6 = 18H/6 = 3H$
Therefore, $3H=54$ and $H=18$

Sol: 10

50 men	40%	25 days
n men	60%	25days

n = 50x(60/40) = 75 men.

Therefore, 75 - 50 = 25 more men are required.

---

6. TIME SPEED DISTANCE & RACE

Distance = speed x time
Distance is directly proportional to speed and time.
Speed is inversely proportional to time.
When speed is increased, time taken will proportionally decrease distance being constant. When speed is decreased, time taken will proportionally increase distance being constant.
Average speed = Total distance/Total time taken.
Distance being constant, when a person travels up and down with different speeds, the average speed of the person is the harmonic mean of the different speeds.
The time taken by a train to cross a pole is equal to the length of the train divided by the speed of the train.
The time taken by a train to cross a platform is equal to the sum of the length of the train and length of the platform divided by the speed of the train.
When two objects move in the same direction the time taken to overtake one another is equal to the sum of the length of the two objects divided by difference of the speeds of the two objects.
When two objects move in opposite direction the time taken to cross each other is equal to the sum of the length of the two objects divided by the sum of the speeds of the two objects.
Note that the relative speed in opposite direction is the sum of the speeds and in the same direction the difference of the speeds.

BOAT AND STREAM

Let the speed of the boat in still water be S_1
And the speed of the stream be S_2
Therefore the speed of the boat in upstream is $S_1 - S_2$; the speed of the boat in downstream is S_1+S_2

RACES (LINEAR AND CIRCULAR)

5 sec or 10m

--|----------

B **A**

In the above diagram, A beats B by 10m or 5 sec. From this we will be able to find the speed of B. B's speed is 2 m/sec. That is B will take 5 sec to cover 10m. Note that A and B start the race simultaneously.

In the above diagram, A gives B a start of 10m or 5 sec. From this, we will be able to find the speed of B. B's speed is 2 m/sec. Here, B starts the race first. A and B reach the finishing point simultaneously.

There are two types of questions in circular tracks.
a) When do they meet for the first time?
b) When do they meet for the first time at the starting point?
Suppose A and B start a race in a circular track. Let A's speed be x m/sec and B's speed be y m/sec. They start the race simultaneously.
For the first question, the time taken $=\frac{l}{|x-y|}$, where l is length of the track in metres.
For the second question, if t_1 and t_2 are the time taken by A and B respectively to reach the starting point after the start then the LCM of t_1 and t_2 will be the answer.

Suppose there are three persons A, B and C run the race.
For the first question, if t_1 and t_2 are the time taken by A and B, & B and C to meet for the first time respectively, then LCM of t_1 and t_2 will be the answer.
For the second question, if t_1, t_2 and t_3 are the time taken by A, B and C respectively to reach the starting point, then LCM of t_1, t_2 and t_3 will be the answer.

Suppose two persons are running in the opposite direction in a circular track then their relative speed will be sum of their speeds. So for the first question, the time taken $=\frac{l}{x+y}$

SOLVED EXAMPLES

1. If a man travels at 30 km/h, he reaches his destination late by10 minutes but if he travels at 42 km/h then he reaches 10 minutes earlier. Therefore, the distance travelled by him is:
Solution:
Let his usual time be t minutes
Therefore, $30(t+10) = 42(t-10) = \text{Distance}$
$30t+300 = 42t-420$
$12t = 720$
$t = 60$ minutes
$D = 30(t+10) = 30(7/6 \text{ hour}) = 35$ km

2. A small aeroplane can travel at 320 km/h in still air. The wind is blowing at a constant speed of 40 km/h. The total time for a journey against the wind is 135 minutes. What will be the time in minutes for the return journey along with the wind?
Solution:
Against the wind the speed is 280 km/h.
Therefore the distance travelled is 280x135 minutes = 280(9/4) = 630 km.
Time taken along with wind is 630/360 = 21/12 hour or 105 minutes.

3. During a journey of 80 km, a train covers first 60 km with a speed of 40 km/h and completes the remaining distance with a speed of 20 km/h. What is the average speed of the train during the whole journey?
Solution:
$60/40 + 20/20 = 5/2$ hours
$80 \div 5/2$; $80(2/5) = 32$ km/h

4. A man makes his upward journey 16 km/h and downward journey 28 km/h. What is his average speed?
Solution:
$(2 \times 16 \times 28) \div (16+28) = 20.36$ km/h

5. Two trains each 120 m in length run in opposite directions with a velocity of 40 m/s and 20 m/s. How long will it take the two trains to cross each other after they meet?
Solution:
$240/60 = 4$ seconds.

6. Two trains starting at the same time from two stations, 200 km apart going in opposite direction, cross each other at a distance of 110 km from one of them. What is the ratio of their speeds?
Solution:
PROBLEM SOLVING TECHNIQUE
$$Distance and speed are in direct proportion. One train covers 110 km and another covers 90 km. Therefore, their speeds will be in the ratio of 110:90 or 11:9.

7. A car can finish a certain journey in 10 hours at the speed of 48 km/h. In order to cover the same distance in 8 hours, the speed of the car must be increased by:
Solution:
PROBLEM SOLVING TECHNIQUE
$$Speed and time are in inverse proportion.
10 hours 48 km/h
8 hours s
=> 10x48 = 8s; s = 60 km/h

8. A train is moving at a speed of 132 km/h. If the length of the train is 110 m, how long will it take to cross a railway platform 165 m long?
Solution:
PROBLEM SOLVING TECHNIQUE
$$To convert km/h to m/s, multiply the number by 5/18.
(110+165)÷132(5/18)
(275x18)/660 = 7.5 sec

9. Two cars start together in the same direction from the same place. The first goes with a uniform speed of 10 km/h. The second car goes at a speed of 8 km/h in the first hour and increases the speed by 1/2 km each succeeding hour. After how many hours will the second car overtake the first car if both cars go non-stop?
Solution:
Let the two cars meet after t hours.
The first car goes with uniform speed. Hence it covers 10t km.
The second car's speed is in AP, where a = 8 and d = .5
Therefore, the distance covered = t/2[16+(t−1).5]
=>10t = t/2[15.5+.5t]
=>20t = 15.5t+.5t^2
=>4.5t = .5t^2
=> t = 9 hours.

10. A plane left 30 minutes later than the scheduled time and in order to reach its destination 1500 km away in time, it has to increase its speed by 250 km/h from its usual speed. Find its usual speed:

Solution:

Let the usual speed of the plane be s and its usual time be t.

st = 1500

1500/s = 1500/(s+250) + 1/2; $s^2+250s-750000 = 0$

(s+1000)(s−750) = 0

s = 750 km/h

11. Two rockets approach each other; one at 42000 mph and the other at 18000 mph. They start 3256 miles apart. How far are they apart 1 minute before impact?

Solution:

PROBLEM SOLVING TECHNIQUE

In 60 minutes, together they would travel 60000 miles.

Therefore, in one minute they would travel 1000 miles.

12. An express train runs at an average speed of 100 km/h stopping for 3 minutes after every 75 km. A local train runs at a speed of 50 km/h stopping for 2 minutes after every 25 km. If the trains began running at the same time, how many kilometres did the local train run in the time it took the express train to run 600 km?

Solution:

Express train will take 6 hours and 21 minutes to cover 600 km.

0	75	150	225	300	375	450	525	600
	3	3	3	3	3	3	3	

See the following table for the local train.

0	25	50	75	100	125	150	175	200	225	250	275
	2	2	2	2	2	2	2	2	2	2	2

The local train cover 275 km in 5 hours and 30 minutes and it takes 22 minutes for the stoppage. In the remaining 29 minutes it can cover 24 and 1/6 km. Therefore, in 6 hours and 21 minutes it covers 299 and 1/6 km.

13. Walking at $3/4^{th}$ of his usual speed a man reaches his office 20 minutes late. Find his usual time.

Solution:

PROBLEM SOLVING TECHNIQUE

$$Speed and time are in inverse proportion
Let his usual speed and time are s and t.
$3/4^{th}$ of s will increase the man's time to $4/3^{rd}$ of t.
The difference in time is $1/3^{rd}$ of t = 20 minutes.
Therefore, t = 60 minutes

14. Dinesh travels 760 km to his home, partly by train and partly by car. He takes 8 hours if he travels 160 km by train and the rest by car. He takes 12 minutes more if he travels 240 km by train and the rest by car. The speeds of the train and the car respectively are:

a) 80 and 100 b) 100 and 80 c) 120 and 100 d) 100 and 120

Solution:
Let the speed of the train be x and speed of the car be y.
160/x +600/y = 8

STRATEGY
Now go from the options and substitute for x and y. Option 'a' is the answer.
Suppose options are not given. Then you have to make the equations and solve them. Those who are interested can look at the following two methods.

Method 1
Let 1/x = a and 1/y = b
160a+600b = 8 or 20a+75b = 1 or 240a+900b = 12
240a+520b = 41/5
240a+900b = 12
380b = 19/5
b = 1/100; => y = 100 and x = 80

Method: 2

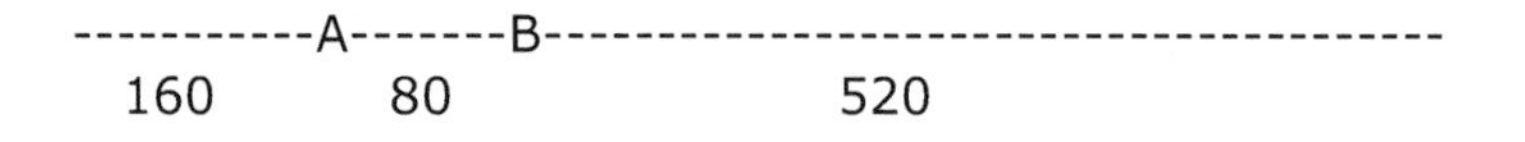

In the first case let the train takes t hours to cover 160 km. Therefore, the car takes 8-t hours to cover 600 km.
In the second case the train takes 3t/2 hours to cover 240 km and the car takes
(8−t)520÷600 hours to cover 520 km.
3t/2 + (8−t)13÷15 = 41/5; => t = 2 hours
Therefore, the speed of the train is 80 and the speed of the car is 100

15. In a flight of 3000 km, an aircraft was slowed down by bad weather. If average speed for the trip was reduced by 100 km/h and the time was increased by one hour, find the original duration of the flight.

Solution:

$3000/(s-100) - 3000/s = 1$ hour

$s^2-100s-300000 = 0$

$(s-600)(s+500) = 0$

Therefore, $s = 600$ km/h and the original duration of the flight is 5 hours.

EXERCISE

1. Points A and B are 70 km apart on a highway. One car starts from A and the another one from B at the same time. If they travel in the same direction, they meet in 7 hours. However, if they travel towards each other, they meet in one hour. The speeds of the two cars are:

2. A journey of 192 km between two cities takes two hours less by a fast train than by a slow train. If the average speed of the slow train is 16 km/h less than that of the fast train, then the average speed of the fast train is:

3. A train can travel 20% faster than a car. Both start from the point A at the same time and reach point B, 75 km away from A at the same time. On the way, however, the train lost about 12.5 minutes while stopping at the stations. The speed of the car is:

4. Excluding stoppages, the speed of a train is 45 km/h and including stoppages, it is 36 km/h. For how many minutes does the train stop per hour?

5. A train leaves station X at 5 am and reaches station Y at 9 am. Another train leaves station Y at 7 am and reaches station X at 10.30 am. At what time do the two trains cross each other?

6. An aeroplane first flew with a speed of 440 km/h and covered a certain distance. It still had to cover 770 km less than what it had already covered, but it flew with a speed of 660 km/h. the average speed for the entire flight was 500 km/h. Find the total distance covered:

7. Train A running at 60 km/h leaves Mumbai for Delhi at 6 pm. Train B running at 90 km/h also leaves Mumbai for Delhi at 9 pm. Train C leaves Delhi for Mumbai 9 pm. If all the three trains meet at the same time between Mumbai and Delhi then what is the speed of train C if distance between Mumbai and Delhi is 1260 km.

8. A boat takes 90 minutes less to travel 36 miles downstream than to travel the same distance upstream. If the speed of the boat in still water is 10 mph, the speed of the stream is:

9. A boat goes 24 km upstream and 28 km downstream in 6 hours. It goes 30 km upstream and 21 km downstream in 6 hours and 30 minutes. The speed of the stream:

10. A boat while going downstream in a river covered a distance of 50 miles at an average speed of 60 mph. While returning because of the water resistance, it took one hour fifteen minutes to cover the same distance. What was the average speed during the whole journey?

11. Two cyclists start on a circular track from a given point but in opposite directions with speeds of 7 m/s and 8 m/s respectively. If the circumference of the circular track is 300 metres after what time will they meet at the starting point?

12. A and B can run 200 m in 22 and 25 seconds respectively. How far is B from the finishing line when A reaches in?

13. A racecourse is 400 metres long. When A and B run a race, A wins by 5 meters. When B and C run over the same course, B wins by 4 metres. When C and D run over it, D wins by16 metres. If A and D run over it, then who would win and by how much?

14. Wheels of diameters of 7 cm and 14 cm start rolling simultaneously from X and Y, which are 1980 cm apart, towards each other in opposite directions. Both of them make same number of revolutions per second. If both of them meet after 10 seconds, the speed of the smaller wheel is :

15. A ship 77 km from the shore springs a leak, which admits to $2\frac{1}{4}$ tonnes of water in $5\frac{1}{2}$ minutes. 92 tonnes of water would sink it. However, the pumps can throw out 12 tonnes of water per hour. Find the average rate of sailing so that the ship may just reach the shore as it begins to sink:

SOLUTIONS

Sol: 1

$70/(s_1-s_2) = 7$
$70/(s_1+s_2) = 1$
$7s_1-7s_2 = 70$ or $s_1-s_2 = 10$
$s_1+s_2 = 70$
Therefore, $s_1 = 40$ and $s_2 = 30$

Sol: 2.

Let the speed of the fast train be s
$192/(s-16) - 192/s = 2$
$s^2-16s-1536 = 0$
$(s-48)(s+32) = 0$
$s = 48$

Sol: 3

If the speed of the car is 100 then the speed of the train will be 120
The ratio of speed of train and car is 6:5
Therefore, the time taken ratio is 5:6
Let the time be 5t and 6t
$6t-5t = 12.5$ minutes
$6t = 75$ minutes.
The car travels 75 km in 75 minutes. Therefore, the speed is 60 km/h.

Sol: 4

PROBLEM SOLVING TECHNIQUE

Let the distance be 36 km
With stoppages it takes 60 minutes
Without stoppages it will take @ 45 km/h, 36/45 = 48 minutes
Therefore, it has taken 12 minutes in stoppages.

Sol: 5

PROBLEM SOLVING TECHNIQUE

The first train takes 4 hours and the second train takes 3½ hours.

Time ratio is 8:7; therefore, speed ratio is 7:8

Let the distance be 28 km and the speeds of the trains as 7 km/h and 8 km/h respectively. At 7 am the first train must have travelled 14 km. The remaining distance is 14 km. The relative speed is 15 km/h.

Therefore, the time taken to meet each other is 14/15 hour or 56 minutes. Hence, they meet at 7.56 am.

Sol: 6

Let the plane travel x km @ 440 km/h. Therefore, it travels x-770 km @ 660 km/h. Let the total time taken be t hours.
$x/440 + (x-770)/660 = t$ hours
Also, $2x-770 = 500t$
$2x = 500t+770$ or $x = 250t+385$
$(250t+385)/440 + (250t-385)/660 = t$
$t = 5\frac{1}{2}$ hours; distance = 500x5.5 = 2750 km.

Sol: 7

Train A starts at 6 pm, whereas train B starts at 9 pm. Therefore, at 9 pm train A must have covered 180 km. Train A and train B will meet after 180/30 = 6 hours. That is at 3 am. They meet at 540 km from Mumbai. Hence train C covers 720 km in 6 hours, or 120 km in 1 hour.

Sol: 8

$36/(10+s_2) + 3/2 = 36/(10-s_2)$
$\Rightarrow s_2 = 2$ mph

Sol: 9

$24/(s_1-s_2) + 28/(s_1+s_2) = 6$ hours
$30/(s_1-s_2) + 21/(s_1+s_2) = 13/2$ hours
let $1/(s_1-s_2) = a$; and $1/(s_1+s_2) = b$
$24a+28b = 6$
$30a+21b = 13/2$; multiply the first with 5 and the second with 4
$b = 1/14$ and $a = 1/6$
$s_1+s_2 = 14$ and $s_1-s_2 = 6$; $\Rightarrow s_2 = 4$ km/h

Sol: 10

Total time taken is 5/6 +5/4 = 25/12 hours.
Total distance = 100 miles.
Therefore, average speed = 100x12/25 = 48 mph.

Sol: 11

The first person comes to the starting point after 300/7 sec.
The second person comes after 300/8 sec.
LCM of 300/7 and 300/8 is 300/1 = 300 sec.
They will meet at the starting point after 300 sec.

Sol: 12
A reaches the finishing point in 22 seconds.
In 22 seconds, B can run 200x22/25 = 176 m.

Sol: 13
When D runs 400 m, C runs 384 m.
When B runs 400 m, C runs 396 m.
Therefore, when C runs 384 m B will run 400x384/396 = 12800/33 m.
When A runs 400 m, B runs 395 m.
Therefore, when B runs 12800/33 m, A will run (400/395)(12800/33) = 392.7 m
Hence, D wins the race by 7.3 m.

Sol: 14
Since both of them make same number of revolutions their speeds will be in the ratio 7:14 or 1:2.
Let their speeds be s and 2s.
Their relative speed in opposite direction is 3s.
Therefore, time taken to meet = 1980/3s = 10; => s = 66 cm per sec.

Sol: 15
The leak admits, in 5½ minutes, 9/4 tonnes of water.
Therefore, in 1 minute it admits 9/22 tonnes of water.
In one minute the pump throws out 12/60 or 1/5 tonnes of water.
The net effect is 9/22 - 1/5 = 23/110 tonnes of water.
The time taken for 92 tonnes of water to get into the ship is 92÷(23/110) minutes
=> 440 minutes or 22/3 hours
The distance to be covered by the ship is 77 km.
Therefore, the required speed is 77/(22/3) = 10.5 km/h.

---

7. EQUATIONS

In this chapter, we will deal with linear equations, quadratic equations and special equations.
Linear equations are of first degree. However, variables can be more than one. In general if there is one variable you need one equation to solve, if there are two variables then you need two equations and so on.

EG: one variable one equation
$3x-8 = 10$; what is the value of x?
Solution:
$3x = 10+8 = 18 \Rightarrow x = 6$

EG: two variables and two equations
$2x+3y = 12$;
$3x+2y = 13$; find the values of x and y.
Solution:
First you have to eliminate one of the variables. In this problem, let us eliminate the variable 'y'. In the first equation the coefficient of y is 3 and in the second it is 2. The technique is that you have to make the coefficients equal. Multiply the first equation with 2 and the second with 3(multiply all the terms).
$4x+6y = 24$ (1)
$9x+6y = 39$ (2)
(2) – (1) gives
$5x = 15; \Rightarrow x = 3$
Now substitute the value of x in one of the given equations.
$2(3)+3y = 12$
$\therefore 3y = 6$
$y = 2$

EG: Three variables and three equations
$x+y+z = 6$
$x-y+z = 2$
$x+y-z = 4$; find the value of x, y and z.
Solution:
The technique is that you take two equations at a time and eliminate any particular variable. You will get two equations in two variables.

Then solve those equations as you have done in the previous example. You will get the
value of two variables. Then substitute the values of those variables in one of the given equations to get the value of the third variable.

$x+y+z = 6$ (1)
$x-y+z = 2$ (2)
(1)+ (2)
$2x+2z = 8$; => $x+z = 4$
$x-y+z = 2$ (2)
$x+y-z = 4$ (3)
(2)+ (3)
$2x = 6$; => $x = 3$
$\therefore z = 1$
$x+y+z = 6$
$3+y+1 = 6$
$\therefore y = 2$

Special equations: Sometimes, in certain problems, you will come across more variables and less number of equations. Such problems cannot be solved using the above techniques. You have to apply some special techniques. Generally, they are not easy to solve. So you are advised to take special attention to such problems. Some examples are given with techniques. Take note of it.
Strategy: when you do not get the technique, go to the options, substitute the values and eliminate the wrong options.
Warning: It might not work out in certain problems!

Quadratic equations

Equations of second degree are called quadratic equations.
The general form is $ax^2+bx+c = 0$; 'a' is the coefficient of x^2, 'b' is the coefficient of x and 'c' is constant. The highest power in the above equation is 2. Hence, it is a second-degree equation. If the power is 3 then it is a cubic equation.
You have a formula to find the roots of the above equation.

$$x = \frac{-b \pm \sqrt{b^2 - 4ac}}{2a}$$

In the above formula, b^2-4ac is called the **'Discriminant'** and is denoted by 'D'. It determines the nature of the roots.
When $D<0$, the roots are complex.
When $D=0$, the roots are rational and equal.
When D is a perfect square, the roots are rational and unequal.
When $D>0$, the roots are irrational and unequal

Therefore, if you know the equation you can find the roots applying the formula. You can also find the roots by factoring the equation. It is illustrated in the following example.

EG: Find the roots of the equation $x^2-5x+6 = 0$.
The technique is that you have to split the coefficient of x in to two parts in such a way that the sum is equal to the coefficient of x and product is equal to the constant.
$-5 = -2, -3$
The given equation can be written as $x^2-2x-3x+6 = 0$
$x(x-2)-3(x-2) = 0$
$(x-2)(x-3) = 0$; => $x-2 = 0$ or $x-3 = 0$
$\therefore$ The roots of the equation are 2 or 3.

If you know the roots of a quadratic equation then you can construct the equation.
If α and β are the two roots of a quadratic equation then the equation is
$(x-\alpha)(x-\beta) = 0$ or $x^2-(\alpha+\beta)x+\alpha\beta = 0$
When you compare the above with the general equation $ax^2+bx+c = 0$, you get two results. $\alpha+\beta = -b/a$; $\alpha\beta = c/a$.

SOLVED EXAMPLES

1. If the length of a certain rectangle decreased by 4 cm and the width increased by 3 cm, a square with the same area as the original rectangle would result. The perimeter of the original rectangle is

Solution:

Let the length and width of the rectangle be x and y respectively.
Therefore, area of rectangle is xy
Now, $(x-4)(y+3)=xy$; also $x-4 = y+3$
$xy+3x-4y-12=xy$; $3x-4y=12$ and $x-y=7$ or $4x-4y=28$
$\Rightarrow x = 16$ and $y = 9$
Perimeter of the rectangle is $2(x+y) = 50$

2. The sum of the digits of a three-digit number is 16. If the ten's digit of the number is 3 times the unit's digit and the unit's digit is one-fourth of the hundredth digit, then what is the number?

Solution:

Let the three digit number be xyz
$x+y+z = 16$; $y = 3z$; $z = 1/4(x)$ or $x = 4z$
Therefore, $4z+3z+z = 16$; $z = 2$
The number is 862.

3. The age of man is 3 times that of his son. 15 years ago, the man was 9 times as old his son. What will be the age of the man after 15 years?

Solution:

Let the age of the man be x and the son be y
$x = 3y$
15 years ago, $x-15 = 9(y-15)$
Therefore, $3y-15 = 9y-135$; $y = 20$ and $x=60$
After 15 years, the age of the man will be $60+15 = 75$.

4. Father is 5 years older than mother and mother's age now is thrice the age of the daughter. The daughter is now 10 years old. What was father's age when the daughter was born?

Solution:

$F = M+5$; $M = 3D$; $D = 10$
Therefore, $M = 30$ and $F = 35$
When the daughter was born, Father's age was 25.

5. In a certain party, there was a bowl of rice for every two guests, a bowl of broth for every three guests and a bowl of meat for every
four guests. If in all there were 65 bowls of food, then how many guests were there in the party?

Solution:

Let the number of guests be x

Therefore, x/2 bowl of rice; x/3 bowl of broth; x/4 bowl of meat

$x/2 + x/3 + x/4 = 65$; $6x/12 + 4x/12 + 3x/12 = 65$

$13x/12 = 65$; $x = 60$

6. There are two examination rooms A and B. If 10 candidates are sent from room A to room B the number of candidates in each room is the same, while if 20 are sent from room B to room A, the number in room A becomes double the number in room B. The number of candidates in each room is respectively:

Solution:

$A-10 = B+10$; $2(B-20) = A+20$

From the first equation $A = B+20$. Substitute in the second equation.

$2B-40 = B+40$; $B = 80$ and $A = 100$

7. A man buys a certain quantity of apples, mangoes, and bananas. If the mangoes were to cost the same as apples, he would have to forgo the bananas to buy the same number of mangoes as he had bought earlier (for the same total amount). The amount spent by him on mangoes and bananas together is 50% more than the amount spent on apples. The total amount spent in the transaction is Rs.140. the number of mangoes bought is the same as the number of bananas. If he wishes to buy the same number of apples as well how much additional amount would have to be spent by him?

Solution:

Let the cost of 1 apple, 1 mango and 1 banana be a, b and c respectively.

Let the number of apples, mangoes and bananas bought be x, y and y.

Therefore, $ax+by+cy = 140$; also $ax+ay = 140$ and $by+cy = 1.5ax$

From 1 and 3 we get, $ax+1.5ax = 140$; $\Rightarrow ax = 56$ and $ay = 84$

To find $ay+by+cy = 84+84 = 168$

Hence, additional amount required is $168-140 = 28$

8. A's age is $1/6^{th}$ of B's age. B's age will be twice of C's age after 10 years. If C's eight birthday was celebrated two years ago, then the present age of A must be:

Solution:

Let A's age be x, then B's age will be 6x

B's age after 10 years is 6x+10; therefore C's age after 10 years is 3x+5.

C's present age = 3x−5; 2 years ago C celebrated his 8th birthday

∴ His present age = 10. Hence, 3x−5 = 10; x = 5

9. A daily wage worker was paid Rs.1700 during a period of 30 days. During this period he was absent for 4 days and was fined Rs.15 per day for absence. He was paid the full salary only for 18 days as he came late on the other days. Those who came late were given only half the salary for that day. What was the total salary paid per month to a worker, who came on time every day and was never absent?

Solution:

During the 30 days, the worker was absent for 4 days, 8 days was late and 18

Days got full salary. Let his full salary for 1 day be x.

Therefore, 18x+4x−60 = 1700; 22x = 1760; x = 80.

Hence, a worker who came on time every day and was never absent will get

30x80 = 2400.

10. 3 chairs and 2 tables cost Rs.700 while 5 chairs and 3 tables cost Rs.1100. What is the cost of 2 chairs and 2 tables?

Solution:

3 chairs and 2 tables = 700 => 6 chairs and 4 tables = 1400

We know 5 chairs and 3 tables = 1100

Subtract 2 from 1 we get. 1 chair and 1 table = 300.

11. $x + \frac{1}{x} = 3$, then the value of $x^2 + \frac{1}{x^2}$ is?

Solution:

$(x+1/x)^2 = x^2+1/x^2+2$

Therefore, $x^2+1/x^2 = 9-2 = 7$

12. $x + \frac{1}{x} = 5$, then the value of $x^3 + \frac{1}{x^3}$ is?

Solution:

$(x+1/x)^3 = x^3+1/x^3+3(x+1/x)$

Therefore, $x^3+1/x^3 = 125-15 = 110$

13. Of the following quadratic equations, which is the one whose roots are 2 and -15

a) $x^2-2x+15=0$ b) $x^2+15x-2=0$ c) $x^2+13x-30=0$
d) $x^2-30=0$

Solution:

If α and β are the roots then the quadratic equation is
$(x-\alpha)(x-\beta) = 0$

Therefore, $(x-2)(x+15)=0$ is the required equation.
$x^2+15x-2x-30 = 0$; $x^2+13x-30 = 0$

EXERCISE

1. The sum of x^2+1 and the reciprocal of x^2-1 is:

2. The solution of the equation $\sqrt{25-x^2} = x-1$ are
   a) 3 and 4   b) 5 and 1   c) −3 and 4   d) 4 and −3

3. Which one of the following is a factor of $x^3-19x+30$
   a) x−2   b) x+2   c) x−1   d) x+1

4. The value of x satisfying the equation $\sqrt{2x+3}+\sqrt{2x-1}=2$ is
   a) 3   b) 2   c) 1   d) 1/2

5. The value of $(1/x^2)+(1/Y^2)$, where $x=2+\sqrt{3}$ and $y=2-\sqrt{3}$, is
   a) 14   b) 12   c) 10   d) 16

6. Two oranges, three bananas and four apples cost Rs.15. Three oranges, two bananas and one apple cost Rs.10. I bought 3 oranges, 3 bananas and 3 apples. How much did I pay?

7. John bought five mangoes and ten oranges together for forty rupees. Subsequently, he returned one mango and got two oranges in exchange. The cost price of an orange would be:

8. A candidate was asked to find $7/8^{th}$ of a positive number. He found $7/18^{th}$ of the same by mistake. If his answer was 770 less than the correct one, what was the original number?

9. My 10-year-old nephew Debu adores chocolates, likes biscuits, and hates apples. One evening I took him to a super market and told him that he could buy as many chocolates as he wanted, but then he should have twice that number of biscuits and finally buy more apples than the total number of chocolates and biscuits. The chocolates cost Rs.1 per piece and apples are twice as expensive; the price of four biscuits matches the price of one apple. Which of the following can possibly be the amount spent by me on Debu's evening snacks?

SOLUTIONS

Sol: 1

$$x^2+1+\frac{1}{x^2-1}=\frac{(x^2+1)(x^2-1)+1}{x^2-1}$$

$$=\frac{x^4-1+1}{x^2-1}=\frac{x^4}{x^2-1}$$

Sol: 2

Squaring both sides we get $25-x^2 = x^2-2x+1$

$2x^2-2x-24 = 0$

$x^2-x-12 = 0$

$x^2-4x+3x-12 = 0$

$(x-4)(x-3) = 0$

Therefore, $x = 4$ or 3

Sol: 3

PROBLEM SOLVING TECHNIQUE

$$Factor theorem

If f(x) is divisible by (x−a) then f(a)=0 and if f(x) is divisible by (x+a) then, f(−a)=0

Now go from the options and substitute the value of x you will get the answer.

When $x=2$; $8-38+30 = 0$

Therefore, $x-2$ is a factor.

Sol: 4

If you go from the options and substitute the value for $x=1/2$, the LHS is equal to 2. Those who are interested in solving the equation can look at the following method.

PROBLEM SOLVING TECHNIQUE

Squaring both sides we get

$4x+2+2\sqrt{(2x+3)(2x-1)} = 4$

$2\sqrt{(2x+3)(2x-1)} = 2-4x$

Once again square both sides.

$32x = 16$; $x = ½$

Sol: 5

$x+y = 4$; $xy = 4-3 = 1$

$(1/x+1/y)^2 = 1/x^2 + 1/y^2 + 2/xy$

$1/x+1/y = (x+y)/xy = 4$

$4^2 = 1/x^2 + 1/y^2 + 2/1$

Therefore, the required answer is 14

Sol: 6

PROBLEM SOLVING TECHNIQUE

You have three variables and only two equations. Therefore, you have to apply some special method as followed below:

$2x+3y+4z = 15$; $3x+2y+1z = 10$

If you add these two equations you will get $5x+5y+5z = 25$

Therefore, $3x+3y+3z = 15$

Sol: 7

Since one mango was exchanged with two oranges, the cost of one mango is equal to the cost of two oranges.

Given that 5 mangoes + 10 oranges = 40

Therefore, 10 oranges + 10 oranges = 40

20 oranges = 40

1 orange = 2

Sol: 8

Let the number be x

($7/8^{th}$ of x) – ($7/18^{th}$ of x) = 770

$(63-28)/72$ of x = 770

$x = 770 \times 72/35 = 22 \times 72 = 1584$

Sol: 9

Let the number of chocolates be x. Then the number of biscuits will be 2x and the number of apples will be greater than 3x.

x	2x	> 3x	Total cost
1	2	4(minimum)	10
2	4	7	18
3	6	10	26
4	8	13	34

Therefore, Rs.34 is a possible amount spent.

8. MENSURATION

1. Area of rectangle = length x breadth

- opposite sides are equal and parallel; all angles are 90°
- diagonals bisect each other
- c)diagonals do not intersect at right angles

2. Area of square = side x side or $(\text{diagonal})^2/2$

- a)all sides are equal; opposite sides are parallel; all angles are 90°
- b)square is a rectangle whose length and breadth are equal
- c)diagonal bisect each other
- d)diagonals intersect at right angles

3. Area of parallelogram = base x height

- a)opposite sides are parallel and equal; opposite angles are equal; sum of adjacent angles equal to 180°
- b)diagonal bisect each other
- c)diagonals do not intersect at right angles
- d)if a rectangle and a parallelogram are drawn on the same base and between the two parallel sides then their areas will be equal
- e)square and rectangle are special types of parallelogram

4. Area of trapezium = $\frac{1}{2}$ x sum of parallel sides x distance between them

- a)only two sides are parallel
- b)diagonals do not bisect each other
- c)diagonals do not intersect at right angles
- d)opposite angles are not equal
- e)only two pairs of adjacent angles are supplementary

5. Area of rhombus = $\frac{1}{2} \times$ product of diagonals

- a)diagonals are not equal
- b)diagonals intersect each other at right angles
- c)diagonals bisect each other
- d)all sides are equal
- e)opposite angles are equal
- f)adjacent angles are supplementary
- g)opposite sides are parallel

6. Area of kite = $\frac{1}{2} \times$ product of diagonals

- a)diagonals are not equal
- b)only the longer diagonal bisect the shorter diagonal
- c)diagonals intersect each other at right angles
- d)angles formed by the longer side and shorter side will be equal
- e)shorter sides are equal; longer sides are equal

7. Area of a quadrilateral = $\frac{1}{2}$ x one of the diagonals (h_1+h_2), h_1 and h_2 are the distances from the opposite vertices to the diagonal.

- a)all sides have different lengths
- b)all angles are different
- c)in a cyclic quadrilateral sum of the opposite angles = 180°

8. Triangles

- a)Area of any triangle = $\frac{1}{2}$x base x height(perpendicular distance from the vertex to the base)
- b)if all sides are different, area= $\sqrt{s(s-a)(s-b)(s-c)}$, where $s = \frac{a+b+c}{2}$; a, b and c are the three sides of the triangle
- c)area of an equilateral triangle = $\frac{\sqrt{3}a^2}{4}$, a is the side of the triangle
- d)area of a triangle = r×s , r is in-radius and s is semi-perimeter
- e)area of a triangle = $\frac{abc}{4R}$; a, b and c are the sides and R is circum-radius
- f)area of a triangle = $\frac{1}{2}$bc sinA; b and c are sides and A is the included angle

9. Circle

- Area of circle = πr^2, r is radius of the circle
- Circumference = $2\pi r$ or πd, d is diameter of the circle
- Area of sector = $\frac{\theta}{360^\circ} \pi r^2$

- Perimeter of sector = $\frac{\theta}{360^\circ} 2\pi r + 2r$
- Length of arc = $\frac{\theta}{360^\circ} 2\pi r$
- Perimeter of segment = $\frac{\theta}{360^\circ} 2\pi r + 2r \sin\frac{\theta}{2}$
- Area of segment = $\frac{\theta}{360^\circ} \pi r^2 - \frac{1}{2} r^2 \sin\theta$
- Area of the portion between two concentric circles = $\pi(R^2 - r^2)$
-

Area of regular hexagon = $6 \text{x} \frac{\sqrt{3}a^2}{4}$ **(there are 6 equilateral triangles)**

SOLID FIGURES

CUBE
Total surface area = $6a^2$
Volume = a^3

CUBOID
Total surface area = $2(lb+bh+lh)$
Volume = lbh

CIRCULAR CYLINDER
Total surface area = $2\pi r(r+h)$
Curved surface area = $2\pi rh$
Volume = πr^2h

CONE
Total surface area = $\pi r(r+l)$, l is slant height
Curved surface area = πrl
Volume = $\frac{1}{3}\pi r^2h$, h is the height of the cone

FRUSTUM

If a cone is cut horizontally by a plane and the smaller cone is removed, then the remaining portion is called a frustum.
Actually, you do not require any separate formula for frustum. You can use the formulae of cone and similar triangle properties to find the area and volume.
If R is the radius of the base, r is the radius of the top, h is the height and l is the slant height of the frustum then,
Lateral surface area = $\pi l(R+r)$
Total surface area = $\pi(R^2+r^2+lR+lr)$
Volume = $\frac{1}{3}\pi h(R^2+Rr+r^2)$

SPHERE
Surface area = $4\pi r^2$
Volume = $\frac{4}{3}\pi r^3$

HEMI-SPHERE

Total surface area = $3\pi r^2$

Curved surface area = $2\pi r^2$

Volume = $\frac{2}{3}\pi r^3$

SPHERICAL SHELL

Total surface area = $4\pi (R^2 - r^2)$

Volume = $\frac{4}{3}\pi (R^3 - r^3)$

SOLID RING

Total surface area = $\pi^2(R^2 - r^2)$

Volume = $\frac{\pi^2}{4}(R-r)^2(R+r)$

SOLVED EXAMPLES

1. PQRS is a diameter of a circle whose radius is r. The lengths of PQ, QR and RS are equal. Semi-circles are drawn on PQ and QS to create the shaded figure below: The perimeter of the shaded figure is:

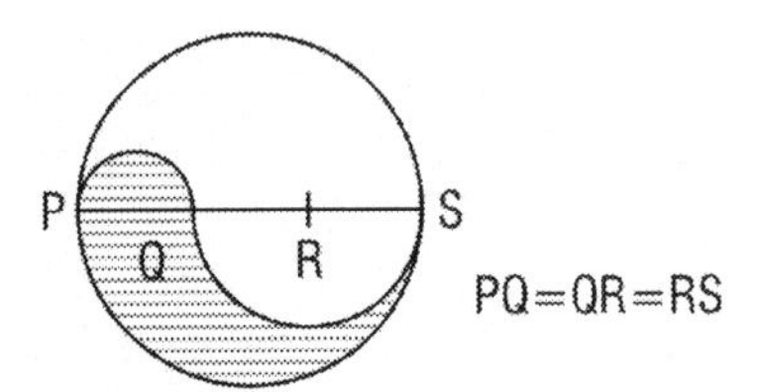

Solution:

PQ = QR = RS = 2r/3

The perimeter of the shaded region is

$\frac{1}{2}(\pi 2r/3) + \frac{1}{2}(\pi 4r/3) + \frac{1}{2}(2\pi r)$

$\pi r/3 + 2\pi r/3 + \pi r = 2\pi r$

2. The floor of a rectangular room is 15 m long and 12 m wide. The room is surrounded by a veranda of width 2 m on all its sides. The area of the veranda is:

Solution:

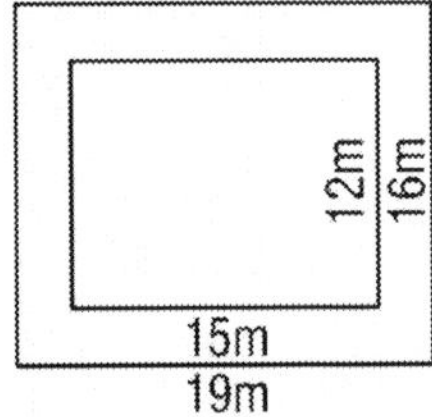

The area of the outer rectangle = (15+4)(12+4) = 304

The area of the inner rectangle = 15x12 = 180

Therefore, the area of the veranda = 124 m^2

3. The ratio of the areas of the inscribed circle to the circumscribed circle of an equilateral triangle is:

Solution:

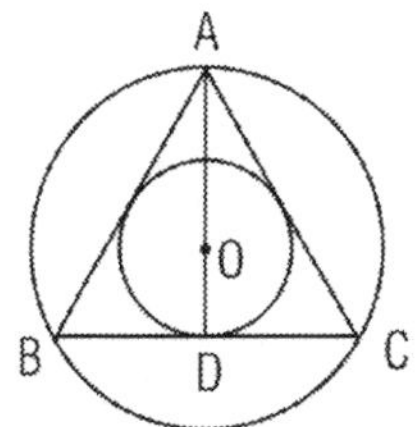

Let ABC be the equilateral triangle. Draw a median AD from the vertex A to the side BC. This will pass through the common centre 'O' of the in-circle and circum-circle.
$$AO/OD = 2/1 (Result)
AO is circum-radius and OD is in-radius
Therefore, the areas of the inscribed and circumscribed circle will be in the ratio of 1:4.

4. The height of a room is 40% of the semi perimeter of its floor. It costs Rs.260 to paper the walls of the room with paper 50 cm. wide @ Rs.2 per metre allowing an area of 15 m^2 for doors and windows. The height of the room is:
Solution:
Let the perimeter of the room be 2(l+b)
Therefore, the height of the room is 0.4(l+b)
The area of the walls = 2h(l+b) or $2\times0.4(l+b)^2$
The net area = $2\times0.4(l+b)^2-15$
Now, the total cost = Rs.260
Cost of the paper for 0.5x1 m^2 = Rs.2
Area of the paper:
Rs.2 → .5 m^2
Rs.260 → A
If you cross multiply you get A = 65 m^2
Therefore, $2\times0.4(l+b)^2-15 = 65$
l+b = 10; h = 4

5. An isosceles triangle is drawn on top of a square whose side is equal to 8 cm. The base of the triangle is equal to the side of the square and its perimeter is 18 cm. Find the area of the figure.

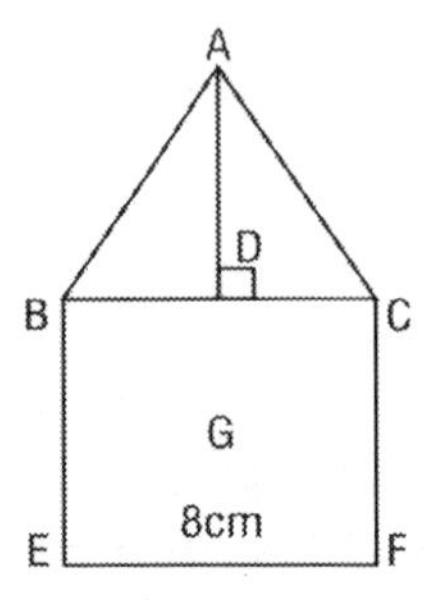

Solution:

In the diagram, AB = AC = 5 cm

$$Draw a vertical line AD that will bisect BC in an isosceles triangle.

Therefore, DC = 4; AD = 3

Area of the triangle = $\frac{1}{2} \times 8 \times 3 = 12cm^2$

Area of the square = $64cm^2$

Therefore, total area = $76cm^2$

6. If the ratio of the areas of two circles is 4:9, the ratio of their circumferences is

Solution:

$$If the ratio of the areas of two circles is a:b, then the ratio of their radii will be $\sqrt{a}$: $\sqrt{b}$, so their circumferences. Therefore, the required ratio is 2:3

7. The ratio of the areas of the in-circle and circum-circle of a square is

Solution:

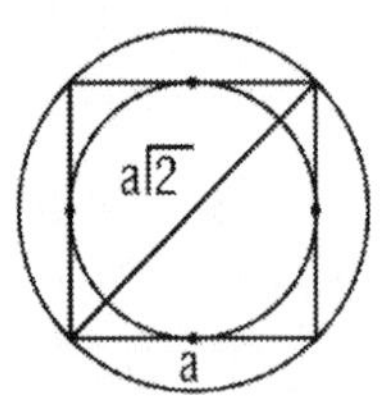

Let the side of the square be a; the length of the diagonal is $\sqrt{2}$ a, and that will be the diameter of the circum-circle. The diameter of the in-circle is the length of the side of the square which is equal to a.

Therefore, the ratio of the diameters are in the ratio of a: $\sqrt{2}$ a.

The ratio of the areas will be the square of the above ratio.

Hence, the required ratio is 1:2

8. If the length of the diagonal of a square and that of the side of another square are both 10 cm, the ratio of the area of the first square to that of the second is

Solution:

Let the side of the first square be a, then diagonal will be $\sqrt{2}$a; a = 10/$\sqrt{2}$

Area of the first square is 50; area of the second square is 100.
Therefore, the required ratio is 1:2

9. If the circumference and the area of a circle are numerically equal, then what is the numerical value of the diameter?

Solution:

$2\pi r = \pi r^2$

Therefore r = 2; diameter = 4

10. The circular park of 20 metre diameter has a circular path just inside the boundary of width 1 metre. The area of the path is

Solution:

Area of the outer circle is 100π
Area of the inner circle is 81π
Therefore, area of the path is 19π.

11. The sides of a rectangular field are in the ratio 3:4 with its area as 7500 m^2. The cost of fencing the field @ 25 paisa per metre is

Solution:

Let the length and breadth of the field be 4x and 3x; area = $12x^2$ = 7500
Therefore, x = 25.
Perimeter = 2(l+b) = 350 metre.
Cost of fencing = 350x.25 = Rs.87.50

12. The sides of a triangle are 6 cm, 11 cm and 15 cm. The radius of its in-circle is:

Solution:

@@Area of the triangle = in-radius x semi-perimeter. Δ = r x s; s = (a+b+c)/2.

r = 1/s$\left(\sqrt{s(s-a)(s-b)(s-c)}\right)$ =1/16$\left(\sqrt{16\times10\times5\times1}\right)$ = 1/16$\left(\sqrt{800}\right)$ = $5\sqrt{2}/4$ cm.

13. A horse is tethered to one corner of a rectangular grassy field 40 m by 24 m with a rope 14 m long. Over how much area of the field can it graze?

Solution:

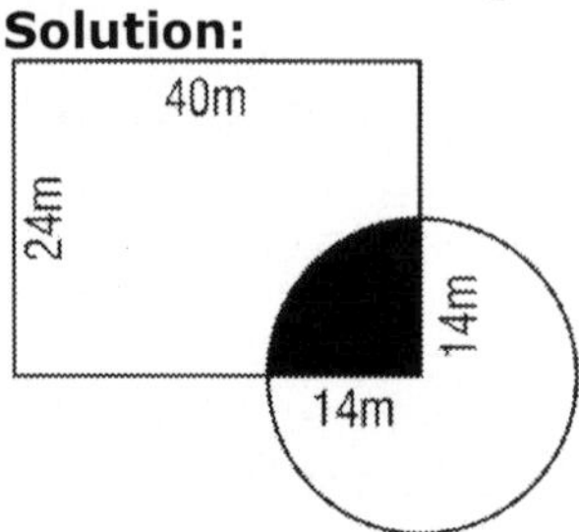

Look at the diagram. The radius of the circle is 14 and the horse can graze $1/4^{th}$ of the area of the circle. $1/4(22/7 \times 14 \times 14) = 154$ m^2

14. A square and an equilateral triangle have the same perimeter. If the diagonal of the square is $12\sqrt{2}$ cm, then the area of the triangle is

Solution:

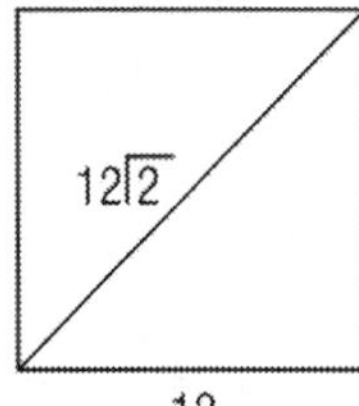

The side of the square will be 12; perimeter = 48; therefore side of the equilateral triangle is 16; area of the equilateral triangle = $\sqrt{3}\,a^2/4$ = $\sqrt{3}$x16x16/4 = $64\sqrt{3}$ cm^2

15. What is the area of the inner equilateral triangle if the side of the outermost square is 'a'

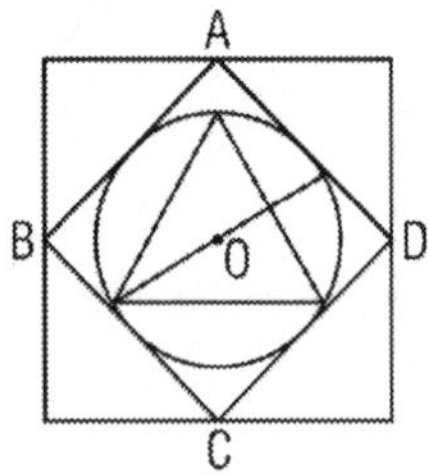

Solution:

In the diagram BD = a, the diagonal of the inner square. Therefore, the side of the inner square is $\frac{a}{\sqrt{2}}$. That is the diameter of the circle. Therefore, the radius the circle is $\frac{a}{2\sqrt{2}}$. The radius of the circle is the circum-radius of the equilateral triangle.

$$The circum-radius is 2/3rd of the altitude of the triangle. If the side of the triangle is x, then altitude is $\frac{\sqrt{3}x}{2}$. Therefore, $\frac{2}{3}\times\frac{\sqrt{3}x}{2} = \frac{a}{2\sqrt{2}}$; $x = \frac{\sqrt{3}a}{2\sqrt{2}}$.

Area of the equilateral triangle is $\sqrt{3}x^2/4$. Substituting the value of x, we get $\frac{3\sqrt{3}a^2}{32}$

EXERCISE

1. The area of the four walls of a room is 120 m^2. The length is twice its breadth. If the height of the room is 4 m, then the area of the floor is:

2. The cross section of a canal is in the form of a trapezium. If the canal top is 10 m wide, the bottom is 6 m wide and the area of the cross section is 72 m^2, then the depth of the canal is:

3. ABCD is a parallelogram. P, Q, R and S are points on sides AB, BC, CD and DA respectively such that AP = DR. If the area of the parallelogram is 16 cm^2, then the area of the quadrilateral PQRS is:
4. Let A be the area of a square inscribed in a circle of radius 'r', and let B be the area of a hexagon inscribed in the same circle. Then B/A equals

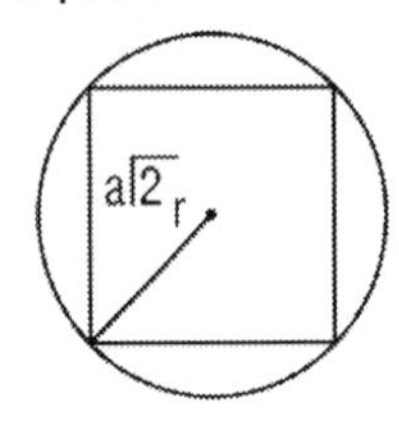

5. In the figure, ABCD is a square with side 10. BFD is an arc of a circle with centre C. BGD is an arc of a circle with centre A. What is the area of the shaded region?

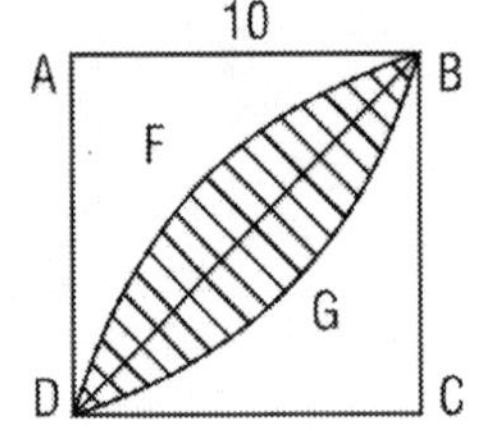

6. Four equal circles are described about the four corners of a square so that each touches two of the others. If each side of the square is 14 cm, then the area enclosed between the circumferences of the circles is:

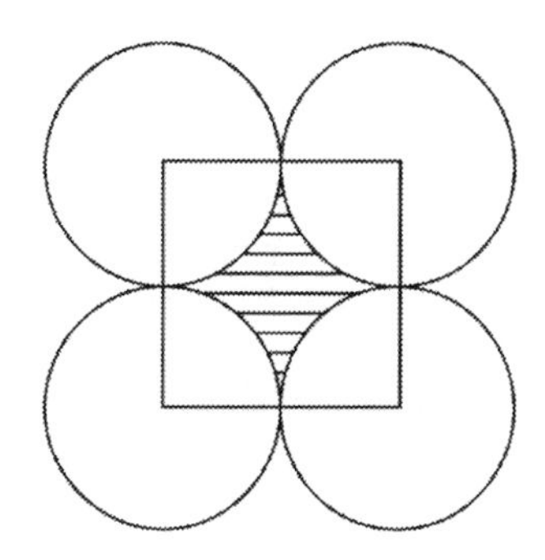

7. If the perimeter of an isosceles right triangle is $(6+3\sqrt{2})$ m, then the area of the triangle is:

8. In the given figure, two circles pass through each other's centre. If the radius of each circle is 2, then what is the perimeter of the region marked B?

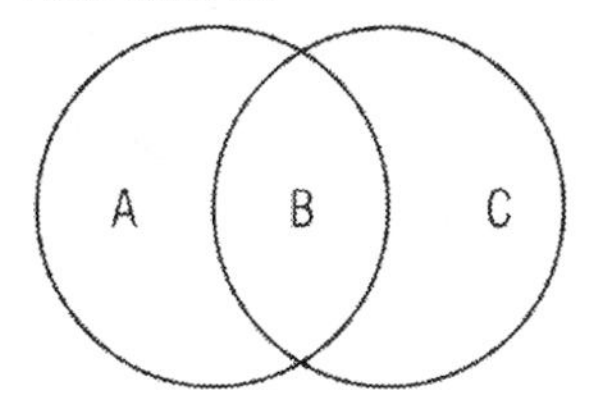

9. If $5\pi/6$ is the measure of each interior angle of a regular convex polygon, then it must be a:

10. The length of an edge of a hollow cube open at one face is $\sqrt{3}$ m. What is the length of the longest pole that it can accommodate?

11. A tank 30 m long 20 m wide and 12 m deep is dug in a field 500 m long and 30 m wide. By how much will the level of the field rise if the earth dug out of the tank is evenly spread over the field?

12. A cylindrical vessel of radius 4 cm contains water. A solid sphere of radius 3 cm is lowered into the water until it is completely immersed. The water level in the vessel will rise by

13. A cylinder is circumscribed about a hemisphere and a cone is inscribed in the cylinder to have its vertex at the centre of one end, and the other end as its base. The volume of the cylinder, hemisphere and the cone are respectively in the ratio:

14. A toy is in the form of a cone mounted on a hemisphere of radius 3.5 cm. The total height of the toy is 15.5 cm. Find the total surface area:

15. The height of a bucket is 45 cm. The radii of the two circular ends are 28 cm and 7 cm respectively. The volume of the bucket is

16. A metallic sheet is of rectangular shape with dimensions 48mx36m. From each of its corners, a square is cut off to make an open box. The volume of the box is x m^3. If the length of the square is 8 m, what is the value of x?

17. The length of the longest rod that can be kept in a room, which is 12m long, 9 m broad and 8 m high, is

18. It is required to fix a pipe such that water flowing through it at a speed of 7 metres per minute fills a tank of capacity 440 cubic metres in 10 minutes. The inner radius of the pipe should be

19. The line AB is 6 metres in length and is tangent to the inner one of the two concentric circles at point C. It is known that the radii of the two circles are integers. The radius of the outer circle is
a) 5 metres b) 4 metres c) 6 metres d) 3 metres

20. ABC forms on equilateral triangle in which B is 2 km from A. A person starts walking from B in a direction parallel to AC and stops when he reaches a point D directly east of C. He, then, reverses direction and walks till he reaches a point E directly south of C. Then D is

a) 3 km east and 1 km north of A
b) 3 km east and $\sqrt{3}$ km north of A
c) $\sqrt{3}$ km east and 1 km south of A
d) $\sqrt{3}$ km west and 3 km north of A

21. In the following figure, PQRS is a square and SR is a tangent to the circle with radius OS. OR intersects the circle at T and OS = TR. What is the ratio of the area of the circle to that of the square?

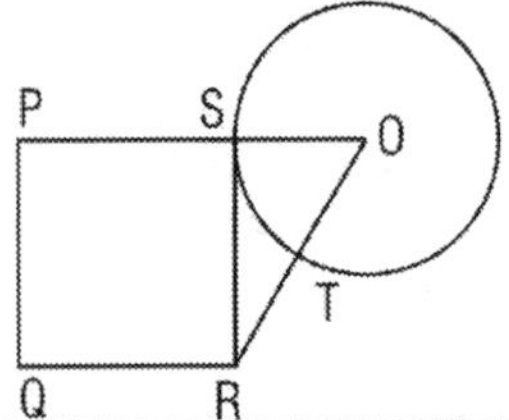

SOLUTIONS

Sol: 1

@@Area of the walls = 2h(l+b) = 2h(3b) = 6hb
h = 4; => A = 24b = 120; b = 5; l = 10;
Area of the floor = 50 m^2

Sol: 2

@@Area of the trapezium is $\frac{1}{2}(a+b)h$; a and b are the two parallel sides and h is the height of the trapezium.
Therefore, h = 9 m.

Sol: 3

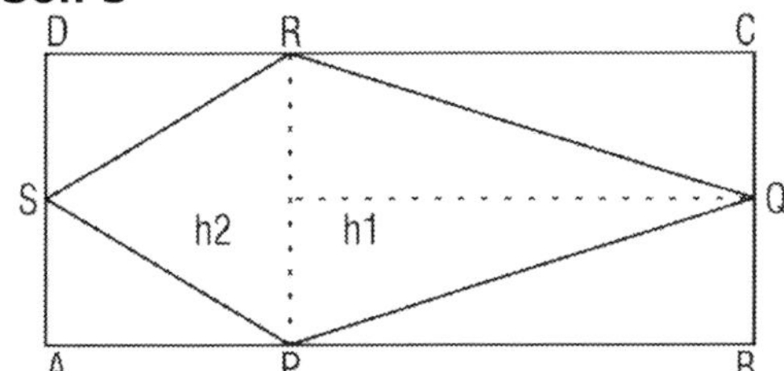

@@Area of the quadrilateral = ½[h1+h2]PR = ½[AB]PR
ABxPR = 16
Therefore, area of the quadrilateral = 8 cm^2

Sol: 4

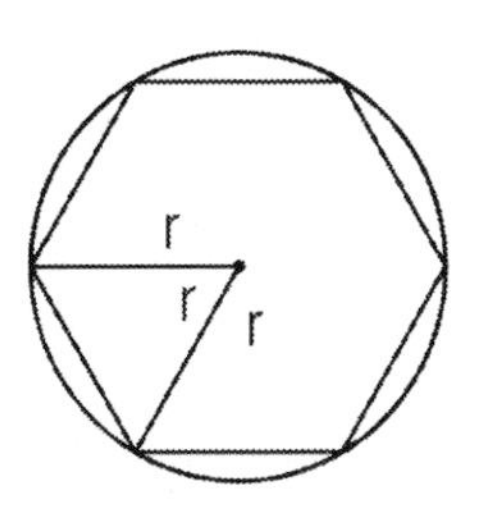

2r will be the diameter of the square. Therefore, area of the square = $2r^2$.

@@The side of the hexagon will be r. therefore, area of the hexagon = $\frac{6\sqrt{3}}{4}r^2$

Hence, B/A = $3\sqrt{3}/4$

Sol: 5

Area of BCDFB is ¼(πx10x10).

Area of the triangle BCD is 50

Therefore, (25π-50) will be half the required area of the shaded region.

Hence, the required area = 50π-100

Sol: 6

Area of the shaded region = area of the square – area of one circle.

Therefore, (14x14)−(πx7x7) = 196−154 = 42 cm^2.

Sol: 7

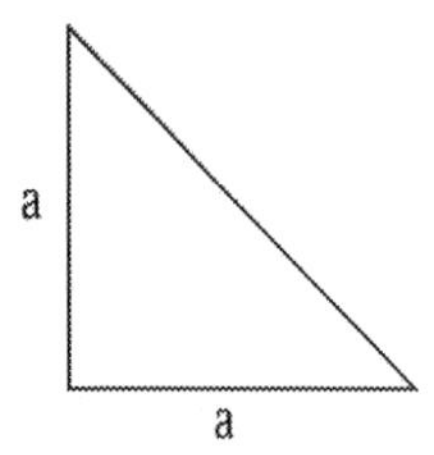

Let the length of the sides, which are equal, be 'a' each. The third side = $\sqrt{2}$a.

Perimeter = 2a+$\sqrt{2}$a = 6+3$\sqrt{2}$; => a = 3.

Therefore, area of the triangle = 4.5 m^2

Sol: 8

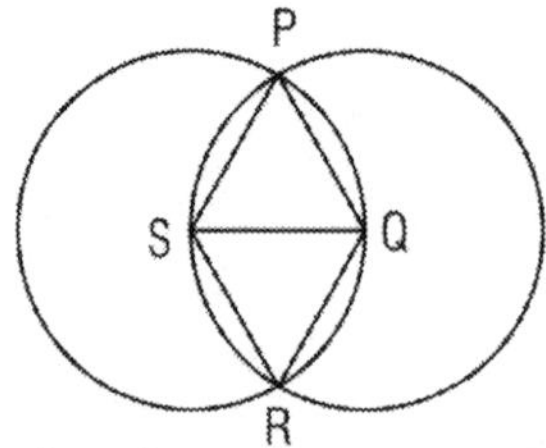

In the diagram PQ = QR = RS = SP = SQ = 2; therefore, triangles PSQ and QSR are equilateral. Angle PSR = 120°. Therefore, length of arc PQR = 1/3x2πr = 4π/3.

Hence, perimeter of the region B is $2 \times 4\pi/3 = 8\pi/3$.

Sol: 9

@@The interior angle of any regular polygon $= \frac{(n-2)180}{n} = \frac{5\pi}{6}$

=5x180/6 = 150.

Which implies n = 12; the polygon is a dodecagon.

Sol: 10

@@The longest diagonal of a cube is $\sqrt{3}$ a, where 'a' is the edge of the cube.

Therefore, required length is 3 m.

Sol: 11

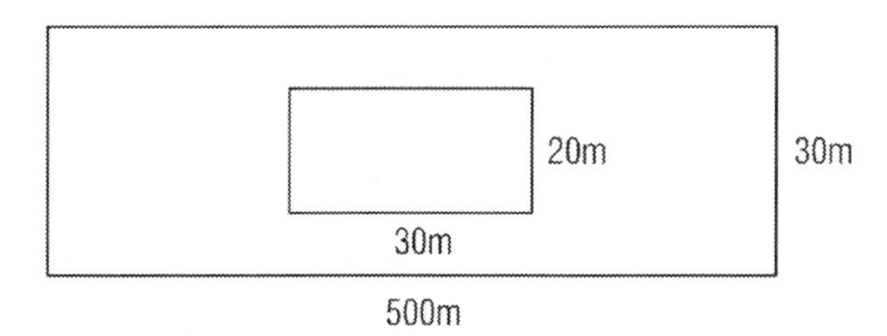

Area of the field is 500x30 = 15000; area of the tank = 30x20 = 600.
Therefore, area of the field where the earth will be filled is 15000-600 = 14400.
Volume of the earth dug out = 30x20x12 = 14400xh (h is the height of the field rise)
Hence, h = ½ m

Sol: 12

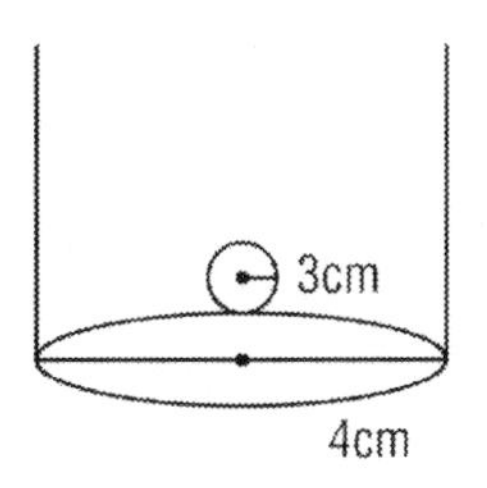

Volume of the sphere = $(4\pi/3)(3 \times 3 \times 3) = 36\pi$
Volume of the water = $\pi \times 4 \times 4 \times h = 36\pi$. Therefore, h =9/4 cm.

Sol: 13

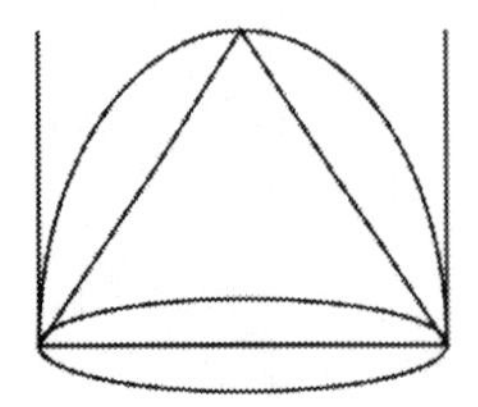

Since the height of the hemisphere is r, the height of the cylinder and the cone is also r. Therefore, πr^3: $4\pi r^3/6$: $\pi r^3/3$

1 : 2/3 : 1/3

3 : 2 : 1

Sol: 14

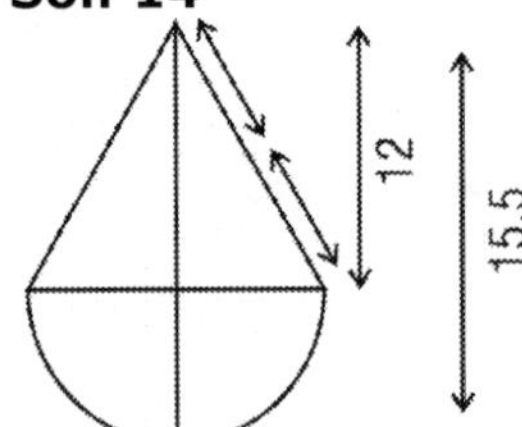

CSA of the cone is (22/7)(3.5xl), where l is the lateral height of the cone.

l = $\sqrt{144+49/4}$ = 12.5; CSA of the hemisphere is 2(22/7)(3.5x3.5)

Total CSA is (22/7)(3.5)(12.5+7) = 214.5 cm^2

Sol: 15

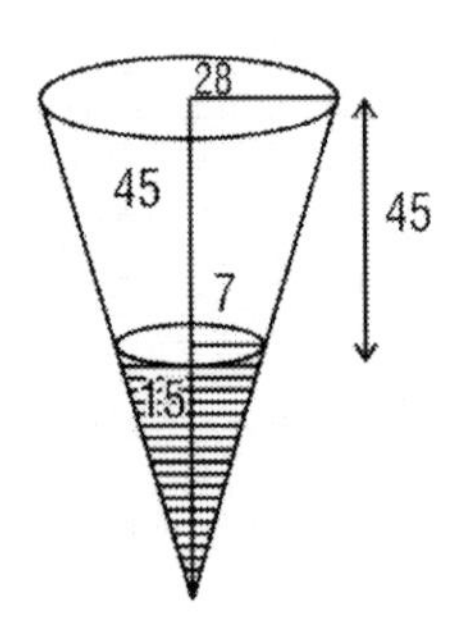

PROBLEM SOLVING TECHNIQUE
Look at the diagram. Volume of the bucket is equal to the difference of the volume of the bigger cone and the smaller cone. The height of the smaller cone is in proportion to the height of the bigger cone (similar triangle properties).

Therefore, 28/7 = (45+h)/h, where 'h' is the height of the smaller cone.
h = 15
Required volume = (1/3)(22/7)[(28x28x60)−(7x7x15)]
V = 48510 cm^2

Sol: 16

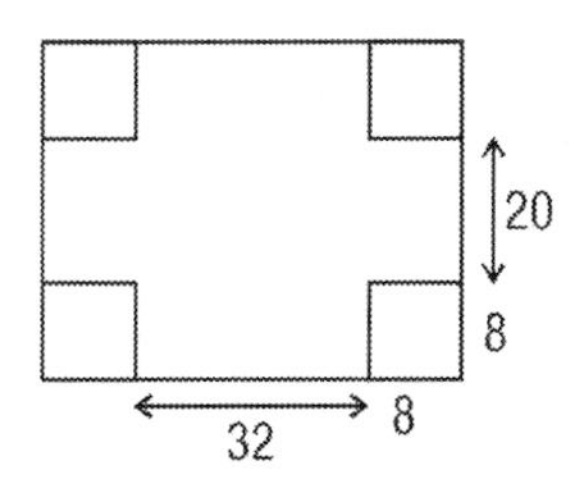

Height of the box = 8 m
Length of the box = 48−16 = 32 m
Breadth of the box = 36−16 = 20 m
Therefore, volume of the box = 32x20x8 = 5120 m^2

Sol: 17
@@The longest rod that can be kept in a room = $\sqrt{l^2+b^2+h^2}$
Therefore, the required length = $\sqrt{144+81+64}$ = $\sqrt{289}$ = 17 m

Sol: 18

@@The volume of water = rate of flow per minute x cross sectional area of the pipe
$7x\pi r^2$ = 44 (in one minute the pipe fills 44 cubic metres)
Therefore, r = $\sqrt{2}$ m.

Sol: 19

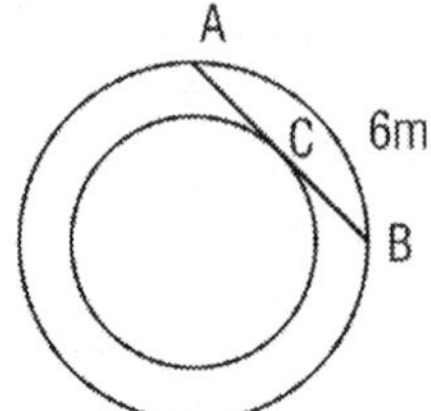

Let centre of the concentric circles be 'O'.
OC is the inner radius and OA is the outer radius.
Triangle OCA is right angled. Therefore, $OC^2+AC^2 = OA^2$
Given, AC = 3 metres and OC, OA are integers, the only option for OA is 5 metres.

Sol: 20

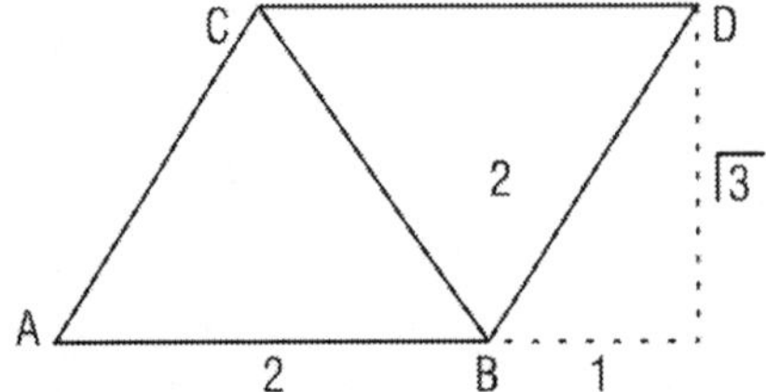

ABCD is a rhombus with each side equals 2 km. Also, BC = 2 km.
Draw a vertical line from D to the line extended from AB. Let it meet at F.
Area of the rhombus is $2x\sqrt{3}x2x2/4 = 2\sqrt{3}$
It is also equal to (base x height) = 2xh = $2\sqrt{3}$.
Therefore, h = $\sqrt{3}$
Now, BFD is a right angled triangle. Therefore, BF = 1 km.
Hence, D is 3 km east and $\sqrt{3}$ km north of A.

Sol: 21
S is on the circle. A tangent is drawn through S and a straight line is drawn from the centre of the circle to S. Therefore, angle OSR is 90°. Let OS=r; therefore, OR=2r and SR=$\sqrt{3}$r. Area of the square = $3r^2$ and area of the circle=πr^2. Hence, the required ratio is $\pi/3$.

---

9. GEOMETRY

LINES

In the figure, PQ and RS are parallel lines. XY is a transverse line. XY intersects the parallel lines PQ and RS at M and N respectively.

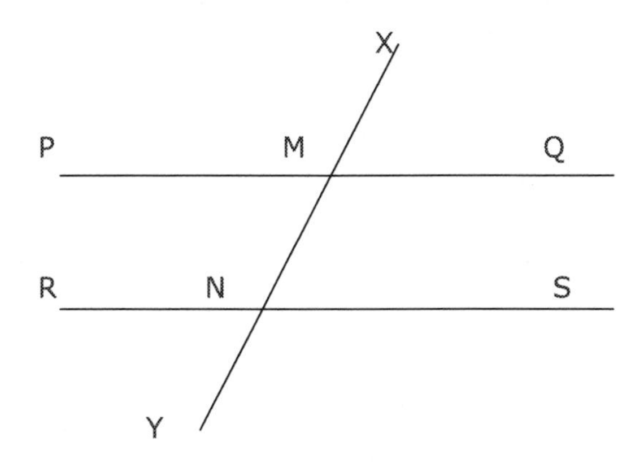

$\angle PMN = \angle MNS$ and $\angle QMN = \angle MNR$ (alternate angles)
$\angle XMQ = \angle MNS$ (Corresponding angles)
$\angle XMP = \angle QMN$ (Vertically opposite angles)
$\angle QMN + \angle MNS = 180^{\circ}$ (Sum of interior angles)

TRIANGLES

Sum of the interior angles of a triangle = 180°
Exterior angle of a triangle = sum of the remote interior angles
Sum of any two sides of a triangle is greater than the third side
The side opposite to the greatest angle will be longest and the side opposite to the least angle will be the shortest

Equilateral triangle: All sides are equal; all angles are equal.
Isosceles triangle: Two sides are equal; the third side is the base; the two equal sides make two angles with the base, which are equal.
Scalene triangle: No two sides are equal; no two angles are equal
Right angled triangle: if one of the angles is 90°

Right angled isosceles triangle: each of the two equal angles = 45° and the third angle = 90°

Median: a line from the opposite vertex to the mid-point of the base
Altitude: a perpendicular line from the opposite vertex to the base
Perpendicular bisector: a perpendicular line passing through the mid-point of any side
Angle bisector: a line, which divides any angle of the triangle into two equal parts

In the diagram below angle ABC<90° (acute triangle), the square of the side opposite to 'angle ABC' is equal to

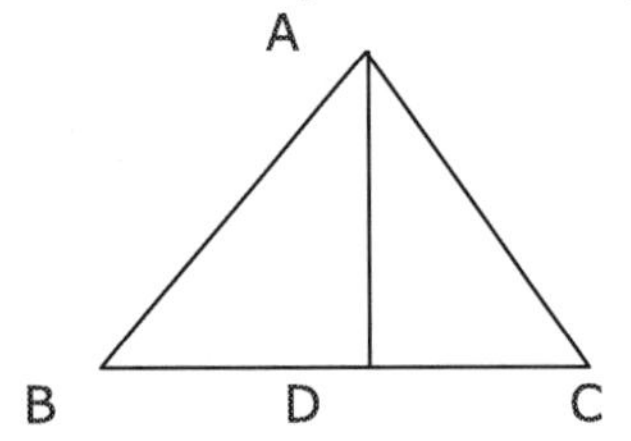

$AC^2=AB^2+BC^2-2BC.BD$

In the diagram below angle ABC>90° (obtuse triangle), the square of the side opposite to angle ABC is equal to

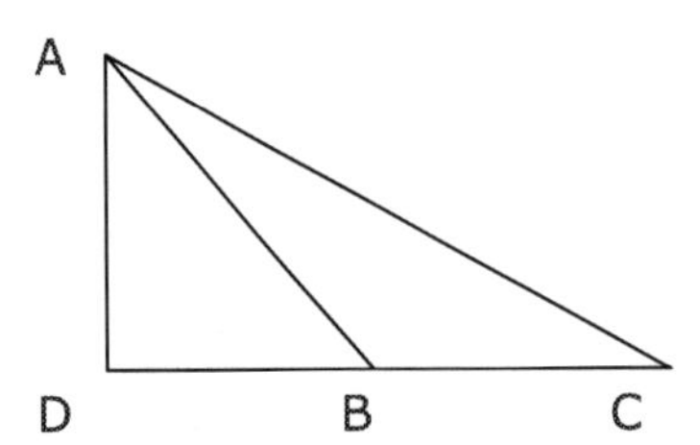

$AC^2=AB^2+BC^2+2BC.BD$

CENTROID

Centroid is a point at which the three medians of a triangle meet.
If A is a vertex, AD is a median and G is the centroid then the ratio of AG to GD is 2/1.
If AD is a median in a triangle ABC then $AB^2+AC^2=2(AD^2+BD^2)$.
In a right angled triangle the median drawn to the hypotenuse will be equal to half of the hypotenuse and the median is the circum-radius of

the triangle, because the hypotenuse is the diameter of the circle circumscribing the triangle.
If AD is a median in a triangle ABC then area of ΔADB= area of ΔADC

CIRCUM-CENTRE

It is a point at which the three perpendicular bisectors of a triangle meet.
In a triangle ABC, if S is the circum-centre then SA=SB=SC=R. R is called the circum-radius. A circle drawn with radius R will circumscribe the triangle ABC.

IN-CENTRE

It is a point at which the three angle bisectors meet inside the triangle. If 'I' is in-centre, the perpendicular lines drawn to the three sides of the triangle from 'I' are all equal. The perpendicular line from I to any side is the in-radius denoted by r. In triangle ABC, angle BIC = $90^\circ + \frac{1}{2}\angle A$.

If AD is an angle bisector of triangle ABC, then $\frac{AB}{AC} = \frac{BD}{DC}$

ORTHOCENTRE

It is a point at which the three altitudes drawn from the three vertices of a triangle meet and is denoted by O.
Note 1: In an isosceles triangle, all the four centres will lie on the same line.
Note 2: In an equilateral triangle, all the four centres will coincide.
Note 3: In an equilateral triangle ABC, if AD is a median and G is the centroid then AG is circum-radius and GD is in-radius. R/r = 2/1. Therefore, $R^2/r^2=4/1$. It is ratio of the area of circum-circle to the area of in-circle.

Note 4: If 'a' is the length of the side of the equilateral triangle ABC then AD will be equal to $\frac{\sqrt{3}a}{2}$ because AD is also the altitude. From this, you will be able to determine the length of circum-radius and in-radius.

SIMILAR TRIANGLES

Two triangles are said to be similar if the three angles of one triangle are equal to the three angles of the second triangle. If a, b and c are the three angles of the first triangle then the three angles of the second triangle will also be a, b and c.
The corresponding sides of the two triangles will be proportional.
If ΔABC is similar to ΔPQR then $\frac{AB}{PQ}=\frac{BC}{QR}=\frac{CA}{RP}$ (the sides opposite to the corresponding angles will be in proportion)

Two triangles will be similar if,
a) All the three angles of one triangle are equal to all the three angles of second triangle (AAA)
b) Two sides of one triangle are proportional to two sides of the second triangle and the included angles are equal (SAS)

If two triangles are similar then
Ratio of corresponding sides = ratio of altitudes = ratio of medians = ratio of in-radii = ratio of circum-radii
Ratio of areas = ratio of square of corresponding sides

In a right angled triangle ABC, right angled at A, if AD is the altitude then all the triangles ABD, CAD and CBA will be similar.
$AD^2=BD.DC$

CONGRUENT TRIANGLES

Two triangles are said to be congruent if they are identical in all respects.
If two triangles are congruent then
a) Corresponding sides are equal
b) Corresponding angles are equal
c) Areas of the two triangles are equal.

Two triangles are congruent if,
a) Corresponding sides are equal (SSS)
b) Two sides and the included angle of one triangle are equal to the two sides and the included angle of the second triangle (SAS)
c) Two angles and one side of a triangle are equal to the two angles and the corresponding side of the second triangle (ASA)
d) The hypotenuse and one side of a right-angled triangle are equal to the hypotenuse and one side of a second right-angled triangle (RHS)

In a triangle ABC, BC is the base and A is the vertex opposite to BC. If PQ is drawn parallel to BC (P is on AB and Q is on AC) then AP/PB = AQ/QC. Also, AB/QP = AC/AQ because APQ and ABC are similar triangles.

POLYGONS

A polygon has more than two sides. Hence, a triangle is also a polygon. There are two types of polygons viz. convex and concave. If all the interior angles are less than 180° then it is convex otherwise it is concave. A regular polygon is a convex polygon in which all the sides are equal and all the interior angles are equal. Therefore, an equilateral triangle is a regular polygon. In a regular polygon, all the lines drawn from the centre to all the vertices will be equal. It implies that the centre of a regular polygon and the centre of the circle circumscribing the polygon are the same.

The sum of the interior angles of regular polygon = $(n-2)180^\circ$ (n is number of sides)

Therefore, the interior angle of a regular polygon = $\frac{(n-2)180^\circ}{n}$

The sum of exterior angles of any regular polygon = 360°

The sum of exterior angles on both directions = 720°

The number of diagonals of any regular polygon = $\frac{n(n-3)}{2}$

CIRCLES

Chord is a line, joining any two points on the circumference of a circle. A perpendicular line drawn from the centre of a circle to the chord will bisect the chord or a line drawn from the mid-point of a chord to the centre of a

circle will be perpendicular to the chord. Two equal chords will be equidistant from the centre of a circle and conversely, if two chords are equidistant from the centre of a circle then the chords are equal. Note that the two chords need not be parallel.

Secant is a line drawn from a point outside the circle to intersect the circle at two points.

If P is a point outside a circle and two secants PAB and PCD are drawn to the circle, the first secant intersects the circle at A and B and the second intersects at C and D then PA.PB=PC.PD

Tangent is a line passing through only one point on the circumference. Only one tangent can be drawn to a given point.

A line joining the centre of a circle and the point at which the tangent is drawn is perpendicular to the tangent.

From a point outside a circle two tangents can be drawn to the circle and they are equal.

PAB is a secant drawn from point P to a circle, which intersects the circle at A, and B. PT is a tangent passing through a point T on the circle. Then, $PA.PB = PT^2$

In the figure, PT is a tangent and PA is a secant. PA divides the circle such that PCA is a minor arc and PBA is a major arc. $\angle APT = \angle PBA$ (**Tangent-secant theorem**)

If two circles are not touching and one is enclosed in the other circle then no common tangent can be drawn.

If two circles are touching internally then only one common tangent can be drawn.

If two circles are touching externally three common tangents can be drawn.

If two circles are intersecting, each other then two common tangents can be drawn.

If two circles are disjoint then four common tangents can be drawn, out of which two are transverse common tangents.

Length of direct common tangent $=\sqrt{d^2-(r_1-r_2)^2}$, d=distance between the centres.
Length of transverse tangent $=\sqrt{d^2-(r_1+r_2)^2}$, d=distance between the centres.

Arc is a segment of a circle. In general, if the length of an arc is less than half of the circumference then it is called a minor arc and if it is more than half of the circumference then it is called a major arc. If AB is an arc and O is the centre of a circle then $\angle AOB$ is the angle subtended by the arc AB with the centre. The sum of angles subtended by a minor arc and a major arc will be 360°.

Angle subtended by a minor arc with the centre of a circle will be twice the angle subtended by the same minor arc with any point on the major arc.
Angle subtended by a minor arc with any of the points on the major arc will be equal.

Diameter is any chord, which passes through the centre of a circle. Diameter is the longest chord of a circle.
Radius: a line joining the centre of a circle and a point on the circumference of the circle

CO-ORDINATE GEOMETRY

Distance between two points (x_1,y_1) and (x_2,y_2) = $\sqrt{(x_1-x_2)^2+(y_1-y_2)^2}$

The point p dividing the line joining the two points (x_1,y_1) and (x_2,y_2) internally in the ratio m:n is, p= $\left(\frac{mx_2+nx_1}{m+n},\frac{my_2+ny_1}{m+n}\right)$

If it is external then p= $\left(\frac{mx_2-nx_1}{m-n},\frac{my_2-ny_1}{m-n}\right)$

The mid-point p of a line joining two points (x_1,y_1) and (x_2,y_2) is $\left(\frac{x_1+x_2}{2},\frac{y_1+y_2}{2}\right)$

If the three vertices of a triangle are given then centroid G= $\left(\frac{x_1+x_2+x_3}{3},\frac{y_1+y_2+y_3}{3}\right)$

If the three vertices of a triangle ABC are given and a, b and c are the lengths of the sides opposite to $\angle A, \angle B and \angle C$ then the indenter I= $\left(\frac{ax_1+bx_2+cx_3}{a+b+c},\frac{ay_1+by_2+cy_3}{a+b+c}\right)$

Slope of a line joining two points is, m= $\frac{y_2-y_1}{x_2-x_1}$, $x_1 \neq x_2$

If two lines are parallel then the slopes will be equal, $m_1=m_2$

If two lines are perpendicular then $m_1.m_2=-1$

Equation of a line:

a) $y - y_1 = m(x - x_1)$, point slope equation
b) $\frac{y - y_1}{x - x_1} = \frac{y_1 - y_2}{x_1 - x_2}$, two points equation
c) $y = mx + c$, slope intercept equation(c is the y-axis intercept)
d) $\frac{x}{a} + \frac{y}{b} = 1$, two intercepts equation(a is x-axis intercept and b is y-axis intercept)
e) $ax + by + c = 0$, general equation of a line

If m_1 and m_2 are slopes of two lines then the angle between the two lines is

$$\tan\theta = \left|\frac{m_1 - m_2}{1 + m_1 m_2}\right|$$

Perpendicular distance from a point (x_1, y_1) to a line $ax+by+c=0$ is $\left|\frac{ax_1 + by_1 + c}{\sqrt{a^2 + b^2}}\right|$

Perpendicular distance between two parallel lines is $\left|\frac{c_1 - c_2}{\sqrt{a^2 + b^2}}\right|$

SOLVED EXAMPLES

1. In a circle of radius 10 cm, a chord is drawn 6 cm from the centre. If a chord half the length of the original chord was drawn, its distance in centimetres from the centre would be:

Solution:

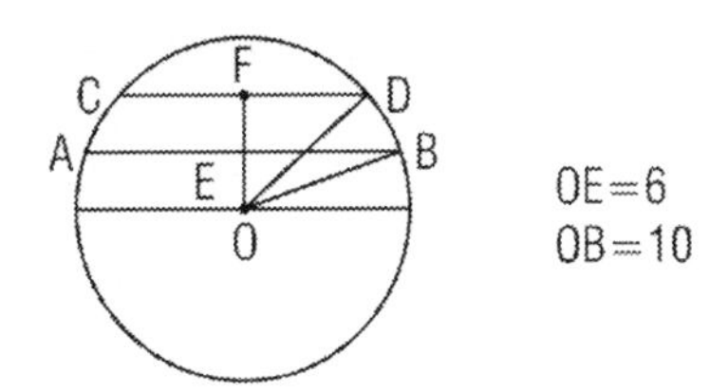

Look at the diagram. Draw a perpendicular line from the centre to the two chords. $$**The perpendicular line from the centre to the chord will bisect the chord.** Let the first chord be AB and the second chord CD. The perpendicular line meets the chords AB and CD at E and F respectively. OE =6; OB = 10; therefore BE = 8. Hence, DF = 4. Again OD = 10 and DF = 4. Therefore, OF = $\sqrt{84}$ cm.

2. The sides of a quadrilateral are extended to make the angles as shown below: what is the value of x?

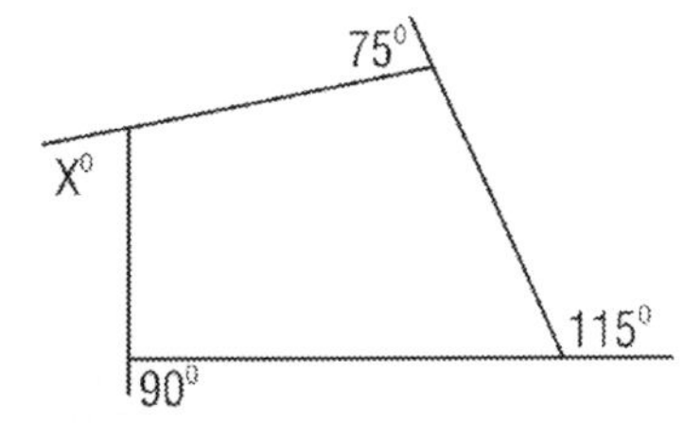

Solution:
@@Sum of the exterior angles of any polygon is 360°. Therefore, x = 80°.

3. What is the area of the region in the Cartesian plane whose points (x, y) satisfy $|x|+|y|+|x+y| \leq$ 2?

Solution:

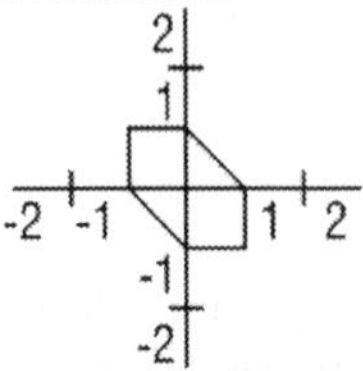

X	1	0	1	-1	-1	0
Y	0	1	-1	1	0	-1

The possible values of x and y are given in the above table. If you draw the graph in the x-y plane you will find the region as given in the diagram. The area of the region is 3 units.

4. The locus of a point equidistant from the two fixed points is:

a) any straight-line bisecting the segment joining the fixed points
b) any straight-line perpendicular to the segment joining the fixed points
c) the straight-line, which is perpendicular bisector of the segment joining the fixed points
d) any straight-line perpendicular to the line joining the fixed points

Solution:

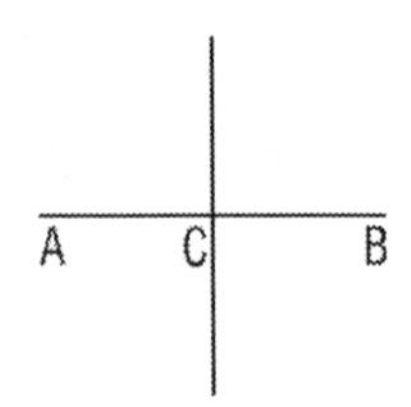

Look at the diagram. Let A and B are the two fixed points.
C is the midpoint of AB.
Draw a perpendicular line through C.
Now, if you take any point on the perpendicular line, the distance from that point to A and B will be equal. Hence, option c is the correct answer.

5. Any cyclic parallelogram having unequal adjacent sides is necessarily a:

a) Square b) Rectangle c) Rhombus d) Trapezium

Solution:
The sides of square and rhombus are equal. Trapezium is not a parallelogram.
Hence the answer is rectangle.

6. The lengths of three sides of a triangle are known. In which of the cases given below, it is impossible to construct a triangle

a) 15 cm, 12 cm, 10 cm
b) 3.6 cm, 4.3 cm, 5.7 cm
c) 17 cm, 12 cm, 6 cm
d) 2.3 cm, 4.4 cm, 6.8 cm

Solution:
$$The sum of any two sides of a triangle is greater than the third side.
In the fourth option 2.3+3.4= 6.7 < 6.8

7. If a parallelogram with area P, a rectangle with area R and a triangle with area T are all constructed on the same base and all have the same altitude, then a false statement is:

a) P=2T b) T=R/2 c) P=R d) none of these

Solution:
$$If a parallelogram and a rectangle are drawn on the same base and with same altitude, then the area of the two will be equal. A triangle with the same base and same altitude will be half the area of the above two. Hence, the answer is none of these.

8. In a circle of radius 17 cm, two parallel chords are drawn on opposite sides of a diameter. The distance between the chords is 23 cm. If length of one chord is 16 cm, then the length of the other is:

Solution:

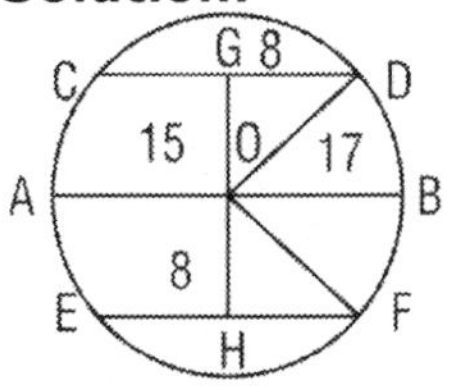

Look at the diagram. OD = 17; GD = 8; therefore, OG = $\sqrt{17^2 - 8^2}$ = 15.
Hence, OH = 8; now, OF = 17; therefore, FH = 15. => EF = 30

9. The three sides of a triangle are given as 4 cm, 3.4 cm, and 2.2 cm. Three circles are drawn with centres at A, B and C in such a way that each circle touches the other two. Then the diameters of these circles would measure (in cm):

Solution:

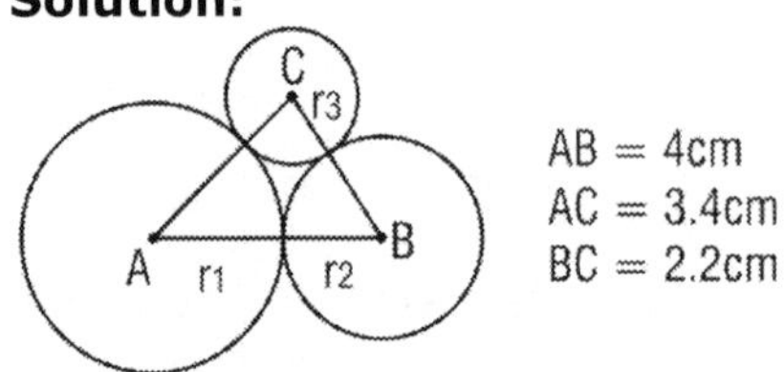

AB = 4cm
AC = 3.4cm
BC = 2.2cm

Look at the diagram. Let r_1, r_2 and r_3 are the radii of the three circles.
Then, $r_1+r_2 = 4$; $r_1+r_3 = 3.4$; $r_2+r_3 = 2.2$; => $r_1+r_2+r_3 = 4.8$
$r_1 = 2.6$; $r_2 = 1.4$; $r_3 = .8$.
Therefore, the diameters will be 5.2, 2.8 and 1.6.

10. In a triangle ABC, $\angle$ A = 90° and D is midpoint of AC. The value of BC^2-BD^2 is equal to:

Solution:

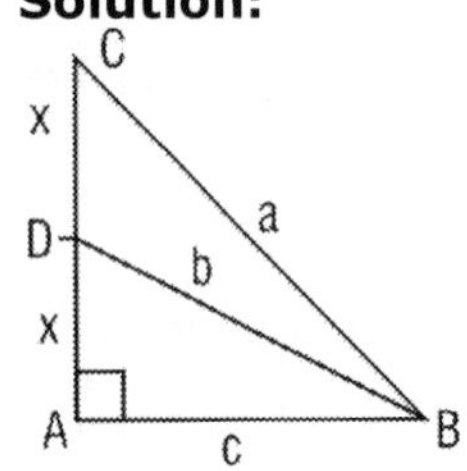

Look at the diagram. Let AD = DC = x; BC = a; BD = b; AB = c.
$a^2 = c^2+4x^2$; $b^2 = c^2+x^2$; => $a^2-b^2 = 3x^2 = 3AD^2$

EXERCISE

1. In the figure given below, O is the centre of the circle. If $\angle OBC = 37^\circ$, then $\angle BAC$ is equal to:

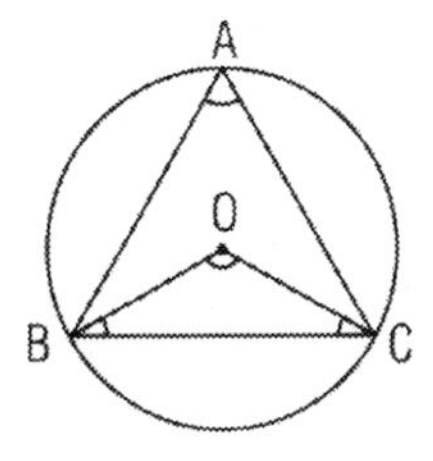

2. In the following figure, find $\angle ADC$.

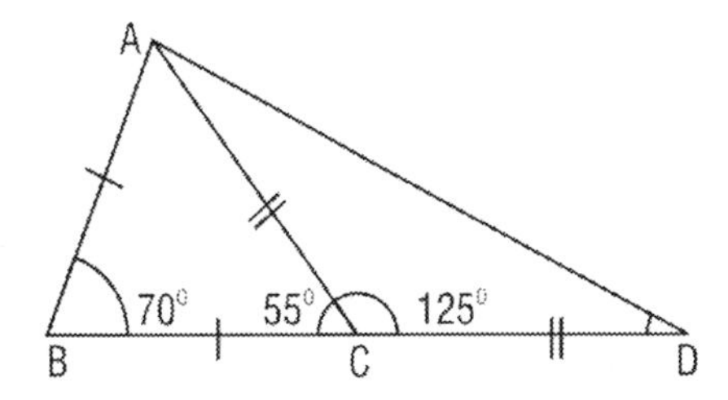

3. How many sides a regular polygon has with the interior angle eight times its exterior angle?

4. The equation of the line through the point of intersection of $3x-y-1 = 0$ and $x-3y+5 = 0$, passing through the point (1, 5) is:

5. The perimeters of two similar triangles ABC and PQR are 36 cm and 24 cm respectively. If PQ = 10 cm, then the length of AB is:

6. In the following figure, PA = 8 cm, PD = 4 cm, CD = 3 cm then AB is equal to:

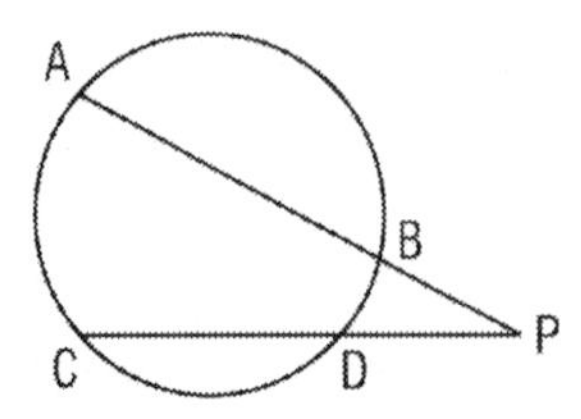

7. The points (0, 8/3), (1, 3), (82, 30) are the vertices of which kind of a triangle.

8. Let ABC be an acute angled triangle and CD be the altitude through C. If AB = 8 cm and CD = 6 cm, then the distance between the mid-point of AD and BC is

9. Two circles of unit radius touch each other and each of them touches internally a circle of radius two units, as shown in the following figure. The radius of the circle, which touches all the three circles, is:

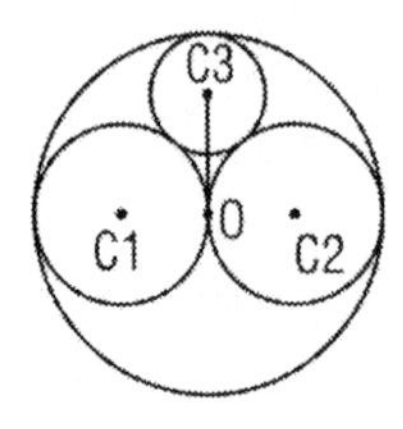

10. In a quadrilateral ABCD, $\angle$ B is 90° and $AD^2 = AB^2+BC^2+CD^2$, then $\angle ACD$ is equal to:

11. The area of a rhombus is 2016 cm^2 and its side is 65 cm. The lengths of the diagonals respectively are:

12. A rhombus OABC is drawn inside a circle whose centre is at O in such a way that the vertices A, B and C are on the circle. If the area of the rhombus is $32\sqrt{3}\,m^2$, then the radius of the circle is:

13. In a triangle ABC, the lengths of the sides AB, AC and BC are 3, 5 and 6 cm respectively. If a point D on BC is drawn such that the line AD bisects the angle A internally, then what is the length of BD?

14. An animal is tethered to point A (see figure below) by a rope. The animal is free to roam anywhere outside the triangle ABC. Neither the

animal nor the rope reaches inside ABC. The angle BAC is 30°. What is the area the animal can cover if the length of the rope is 12 m?

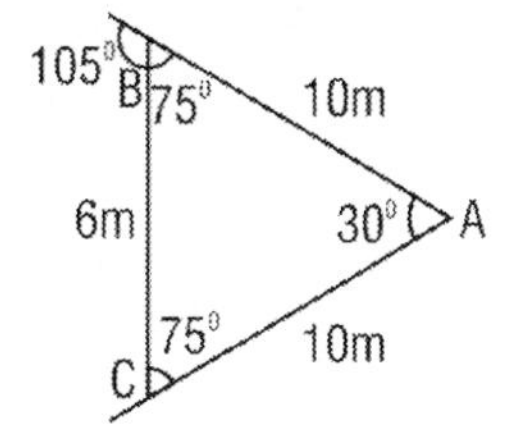

SOLUTIONS

Sol: 1

Look at the diagram. ∠OBC = 37°(given). OB = OC is radius of the circle. Therefore, ∠OCB = 37°. Hence, ∠BOC = 106°. ∠BAC = ½(∠BOC). Therefore, ∠BAC = 53°

Sol: 2

Look at the figure. AB = BC; therefore ∠BAC = ∠BCA = 55°. Therefore, ∠ACD = 125°(linear pair); AC = CD; therefore ∠ADC = 55/2 = 27.5°

Sol: 3

Let 'n' be the number of sides of the regular polygon.
$$Interior angle of a regular polygon is (n−2)180/n.
Exterior angle of a regular polygon is 360/n.
Therefore, (n−2)180/n = 8(360/n).
Hence, n = 18.

Sol: 4

@@If a line is passing through two points (x_1, y_1) and (x_2, y_2) then the equation of the line is ($y-y_1$)(x_1-x_2) = ($x-x_1$)(y_1-y_2).
3x−y = 1; x−3y = −5. Solving the equations we get x = 1; y = 2.
The two lines will pass through the point (1, 2).
Now, you have two points (1, 5) and (1, 2).
(y−5)(1−1) = (x−1)(5−2)
x = 1 is the required equation.

Sol: 5

PROBLEM SOLVING TECHNIQUE
$$**If ABC and PQR are two similar triangles then AB/PQ=BC/QR=AC/PR.**
If ratios are in continued proportion then the ratio of sum of the numerators to the sum of denominators is equal to each of the ratios.
Hence, AB/10 = 36/24; => AB = 15

Sol: 6

Look at the diagram. PA and PC are two secants meeting at P.
$$Then, PAxPB = PCxPD (property).

Therefore, 8xPB = 7X4; => PB = 3.5
Hence AB = 8-3.5 = 4.5

Sol: 7

@@If (x_1, y_1) and (x_2, y_2) are two points, the slope of the line joining the two points is $(x_1-x_2)/(y_1-y_2)$

The slopes of the three lines formed by taking two points at a time are all equal.

Therefore, the points are on a straight line.

Sol: 8

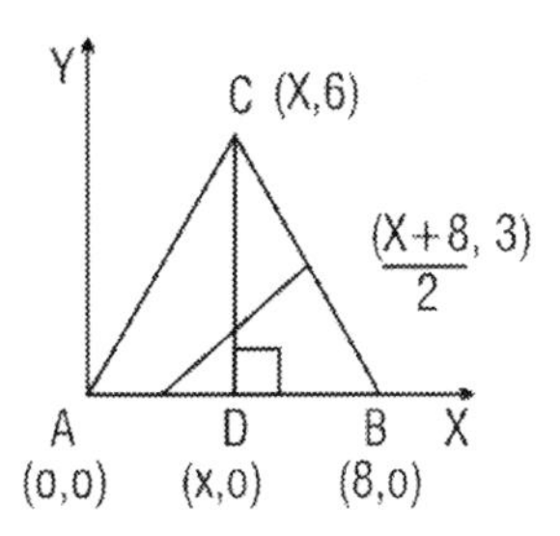

Look at the diagram. You have to use x-y plane to solve this problem. Let the co-ordinates of A be at the origin. Since the length of AB is 8 cm, the co-ordinates of B will be (8, 0). We do not know the co-ordinates of D, which
is on the line segment AB. Let it be (x, 0). The length of CD is 6 cm. therefore, the co-ordinates of C will be (x, 6). Using mid-point formula, the mid-point of AD will be (x/2, 0) and the mid-point of CB will be ((x+8)/2, 3). Now, use the shortest distance between any two points' formula to get the answer. The answer is 5.

Sol: 9

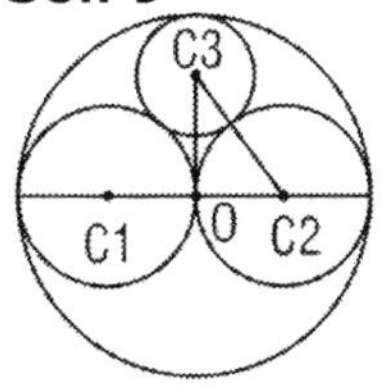

Look at the diagram. Let the radius of the smallest circle be r.
$C_2C_3 = 1+r$; $OC_2 = 1$; $OC_3 = 2-r$ (radius of the biggest circle – radius of the smallest circle). By symmetric nature of the figure C_3OC_2 is a right angled triangle. Therefore, $(1+r)^2 = (2-r)^2 + 1^2$
Solving we get r = 2/3

Sol: 10

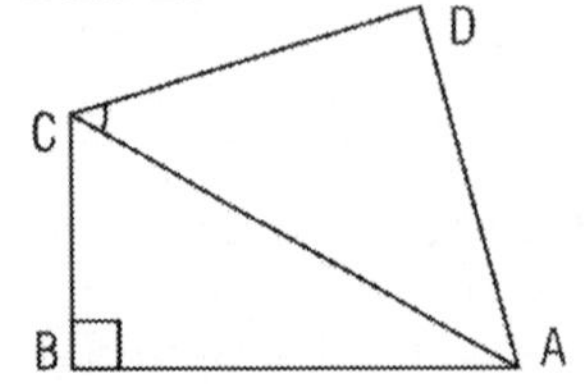

Look at the diagram. $AC^2 = AB^2+BC^2$. Therefore, $AD^2 = AC^2+CD^2$.
It implies $\angle ACD = 90^\circ$

Sol: 11

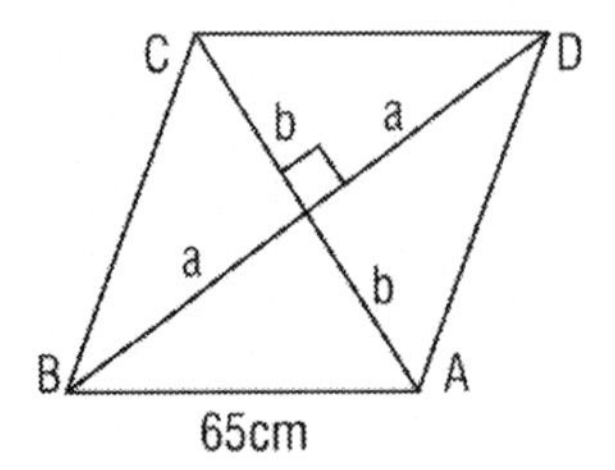

Look at the diagram. The diagonals of a rhombus will be $\perp$r to each other.
Therefore, $a^2+b^2 = 65^2$; also, area of one triangle $ab/2 = 2016/4$.
$ab = 1008$; apply $(a+b)^2$ and $(a-b)^2$ formula to find $a+b$ and $a-b$. Once you find that $a = 63$ and $b = 16$. Hence, the lengths of the diagonals are 126 and 32 respectively.

Sol: 12

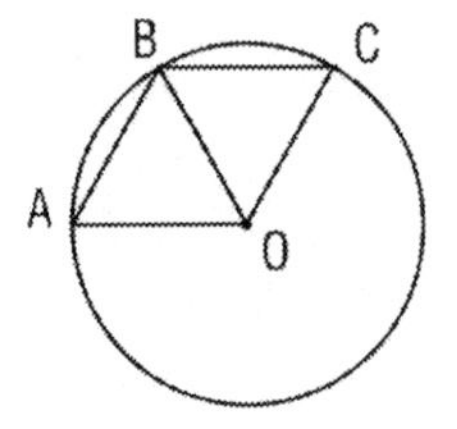

Look at the diagram. OA = OB = OC = r; $\therefore$AB = BC = r.
Area of the rhombus is sum of the two equilateral triangles.

$\frac{2\sqrt{3}r^2}{4} = 32\sqrt{3}$; => r = 8

Sol: 13

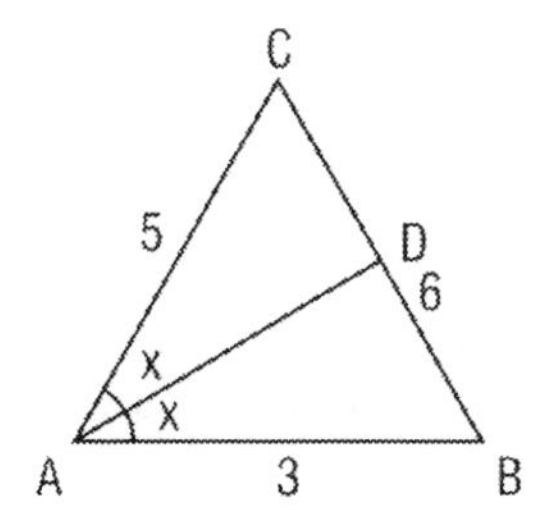

$$Look at the diagram. If AD is bisector angle of A then CD/CA = BD/BA.
Let BD = x then CD = 6-x.
(6-x)/5 = x/3
Solving x = 2.25.

Sol: 14

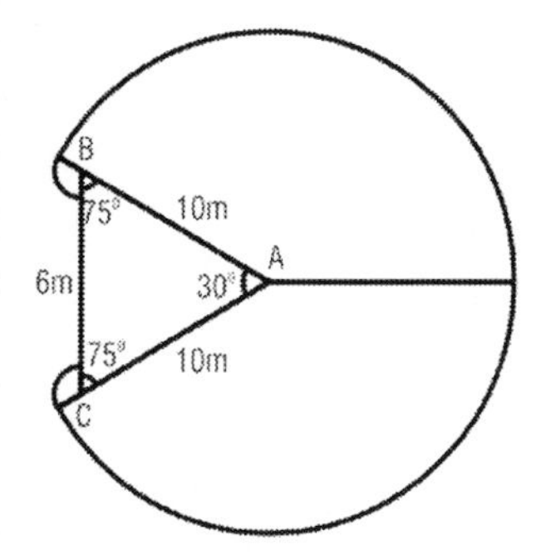

Look at the diagram. The animal can roam three regions marked as A, B and C.
Area of A = (330/360)πx12x12 = 132π.
Area of B and C = 2x(105/360)πx2x2 = 7π/3.
Therefore, the total area = (403/3)π

---

10. TRIGONOMETRY

It is a branch of mathematics, which concerns the relationship of sides and angles of triangles.

In the figure, the side opposite to the angle θ is called opposite side, the side opposite to the right angle is called hypotenuse and the third side is called adjacent side.

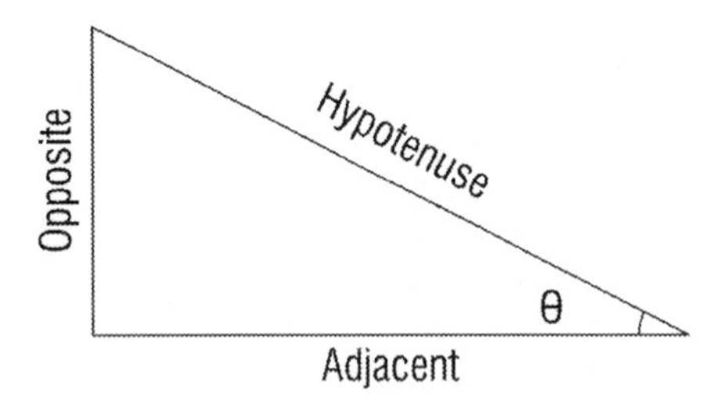

Relationship between degree and radian

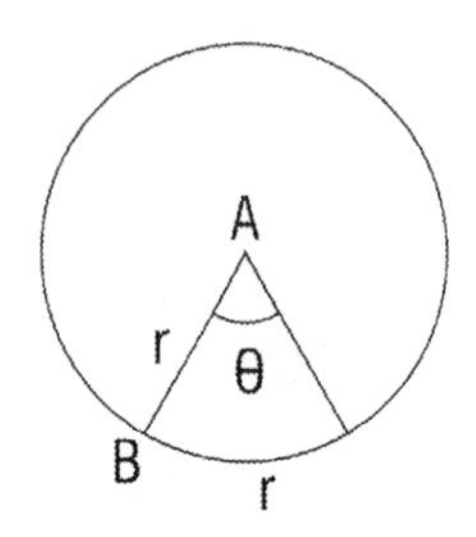

In the figure, A is the centre of the circle and AB is the radius of the circle with length r. If AB is rotated through an arc length of l which makes an angle θ radian at the centre then $\theta^c = l/r$.
It implies, when the arc length is r, $\theta^c = 1$ radian
If AB is rotated through the full length of the circumference then $\theta^c = 2\pi r/r = 2\pi$ radians $= 360°$

The following are the ratios of the sides with respect to the $\angle\theta$.
$\sin\theta =$ opposite/hypotenuse
$\cos\theta =$ adjacent/hypotenuse

$\tan\theta$ = opposite/adjacent or $\sin\theta/\cos\theta$
$\text{cosec}\,\theta$ = hypotenuse/opposite or $1/\sin\theta$
$\sec\theta$ =hypotenuse/adjacent or $1/\cos\theta$
$\cot\theta$ =adjacent/opposite or $1/\tan\theta$

Sign of T-Ratios:

1st quadrant : All are positive
2nd quadrant: sin and cosec are positive
3rd quadrant: tan and cot are positive
4th quadrant: cos and sec are positive
Remember: all silver tea cups

The following table shows values of certain standard angles:

	30	45	60	90	0
Sin	1/2	$1/\sqrt{2}$	$\sqrt{3}/2$	1	0
Cos	$\sqrt{3}/2$	$1/\sqrt{2}$	½	0	1
Tan	$1/\sqrt{3}$	1	$\sqrt{3}$	ND	0
Cosec	2	$\sqrt{2}$	$2/\sqrt{3}$	1	ND
Sec	$2/\sqrt{3}$	$\sqrt{2}$	2	ND	1
Cot	$\sqrt{3}$	1	$1/\sqrt{3}$	0	ND

The following are certain important results:

$\sin^2\theta+\cos^2\theta=1$
$1+\tan^2\theta=\sec^2\theta$
$1+\cot^2\theta=\text{cosec}^2\theta$
$\sin(-\theta)= -\sin\theta$
$\cos(-\theta)= \cos\theta$
$\sin(90-\theta)=\cos\theta$
$\cos(90-\theta)= \sin\theta$
$\sin(90+\theta)= \cos\theta$
$\cos(90+\theta)= -\sin\theta$
$\sin 2a=2\sin a.\cos a$

$\cos 2a = \cos^2 a - \sin^2 a$

$\tan 2a = \frac{2\tan a}{1-\tan^2 a}$

$\cos 2a = \frac{1-\tan^2 a}{1+\tan^2 a}$

$\sin 2a = \frac{2\tan a}{1+\tan^2 a}$

$\sin(a+b) = \sin a.\cos b + \cos a.\sin b$

$\sin(a-b) = \sin a\cos b - \cos a\sin b$

$\cos(a+b) = \cos a\cos b - \sin a\sin b$

$\cos(a-b) = \cos a\cos b + \sin a\sin b$

$\tan(a+b) = \frac{\tan a + \tan b}{1 - \tan a \tan b}$

$\tan(a-b) = \frac{\tan a - \tan b}{1 + \tan a \tan b}$

$\sin 15 = \frac{\sqrt{3}-1}{2\sqrt{2}}$

$\cos 15 = \frac{\sqrt{3}+1}{2\sqrt{2}}$

$\sin 18 = \frac{\sqrt{5}-1}{4}$

$\cos 18 = \frac{\sqrt{10+2\sqrt{5}}}{4}$

$\sin 36 = \frac{\sqrt{10-2\sqrt{5}}}{4}$

$\cos 36 = \frac{\sqrt{5}+1}{4}$

$\sin 22.5 = \frac{\sqrt{2-\sqrt{2}}}{2}$

$\cos 22.5 = \frac{\sqrt{2+\sqrt{2}}}{2}$

SOLVED EXAMPLES

1. A ladder reaches a window, which is 12 m above the ground, and is on one side of the street. Keeping its foot at the same point, the ladder is turned to the other side of the street to reach a window 9 m high. Find the width of the street, if the length of the ladder is 15m.

Solution:

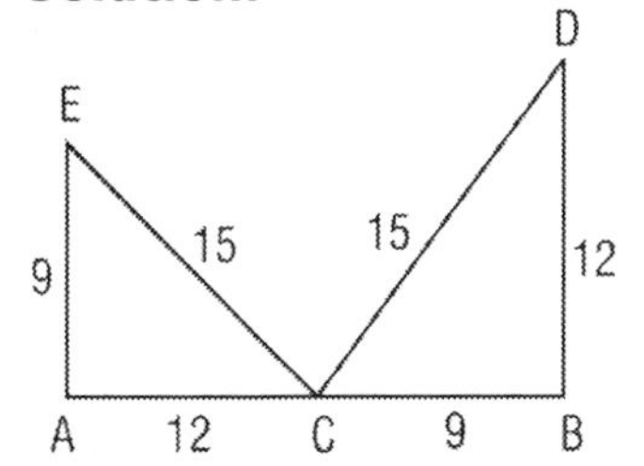

$BC^2 = 15^2 - 12^2$; $\Rightarrow BC = 9$

$AC^2 = 15^2 - 9^2$; $\Rightarrow AC = 12$

Therefore, width of the street is 21.

2. A tree breaks due to storm and the broken part bends so that the top of the tree first touches the ground making an angle of 30∘ with the horizontal. The distance from the foot of the tree to the point where the top touches the ground is 10 m. The height of the tree is:

Solution:

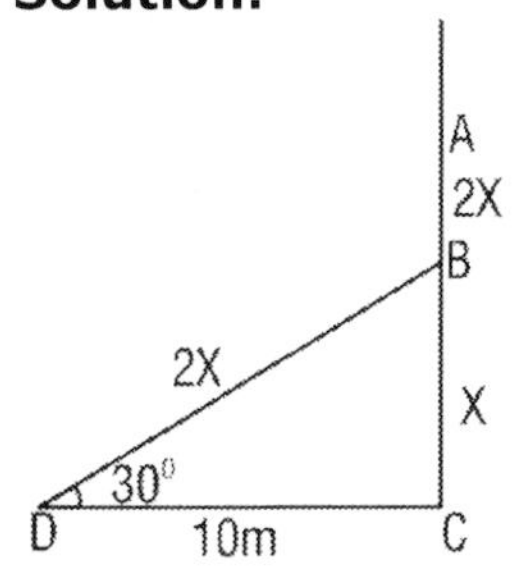

$\sin 30° = 1/2 = BC/BD$

Let BC = x then BD = AB = 2x

$4x^2 - x^2 = 100$; $\Rightarrow x = 10/\sqrt{3}$

Therefore, $3x = 10\sqrt{3}$ m.

3. Find the angle of elevation of the sun when the length of the shadow of a pole is $\sqrt{3}$ times than the height of the pole.
Solution:

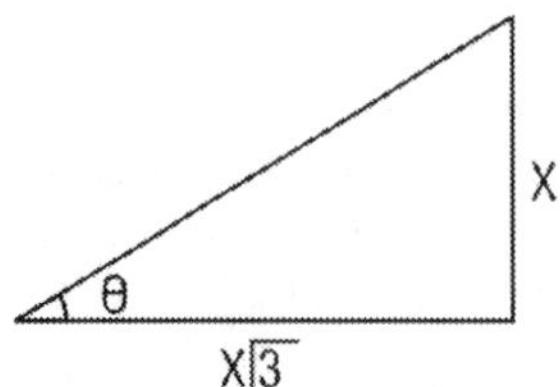

Let the height of the pole be x, then the length of the shadow will be $\sqrt{3}$x.

$\tan\theta = 1/\sqrt{3}$; => $\theta = 30^\circ$

4. If 0<x≤ π /2, then sin x + cosec x ≥

a) 0 b) 1 c) 2 d) none of these

Solution:

When x = 0, the given expression will be equal to ∞(cosec x = 1/sin x)

When x = $\pi/2$, the given expression will be equal to 2 (sin $\pi/2$ = 1)

Therefore, the given expression will be greater than or equal to 2.

5. The angles of elevation of an artificial satellite measured from two earth stations are 30° and 60° respectively. If the distance between the earth stations is 4000 km, then the height of the satellite is:
Solution:

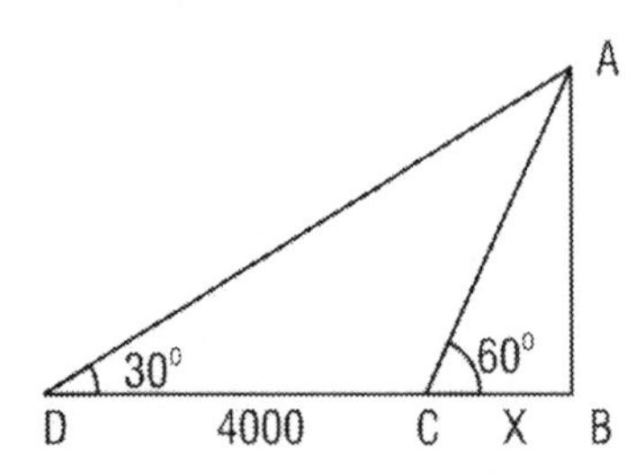

$\tan 60^\circ = \sqrt{3}$ =AB/BC; =>AB = $\sqrt{3}$ BC

$\tan 30^\circ = 1/\sqrt{3}$ =AB/(4000+BC); => 3 BC = 4000+BC

BC = 2000 and AB = $2000\sqrt{3}$.

EXERCISE

1. The angles of elevation of the top of a tower from two points at distances m and n metres are complementary. If the two points and the base of the tower are on the same straight line, then the height of the tower is:

2. If $\sin\theta+\cos\theta = a$ and $\sec\theta+\mathrm{cosec}\theta= b$. then the value of $b(a^2-1)$ is equal to:

3. If $7\mathrm{cosec}\theta$ - $3\cot\theta = 7$, then the value of $7\cot\theta$ - $3\mathrm{cosce}\theta$ is equal to

4. If $a\sec\theta+b\tan\theta = 1$ and $a^2\sec^2\theta-b^2\tan^2\theta = 5$, then $a^2b^2+4a^2$ is equal to

5. From the top of a light house 60 m high with its base at the sea level, the angle of depression of a boat is 30°. The distance of the boat from the foot of the light house is:

6. A 25 m ladder is placed against a vertical wall of a building. The foot of the ladder is 7 m from the base of the building. If the top of the ladder slips 4 m, then the foot of the ladder will slide:

7. From the top of the building 60 m high, the angles of depression of the top and bottom of a tower are observed to be 30° and 60°. Find the height of the tower?

8. If the elevation of the sun changes from 30° to 60°, then the difference between the lengths of shadows of a pole 15 m high at these two elevations of the sun is:

SOLUTIONS

Sol: 1

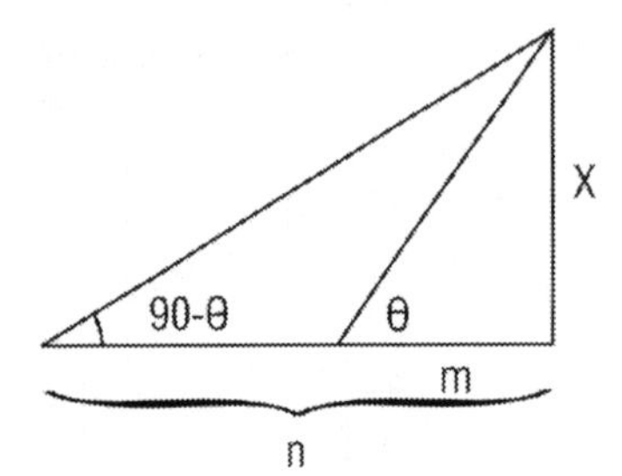

$\tan\theta = x/m$; $\tan(90-\theta) = x/n = \cot\theta$
$\tan\theta.\cot\theta = 1$; $\Rightarrow x^2/mn = 1$; $x = \sqrt{mn}$

Sol: 2
$a^2 = \sin^2\theta + \cos^2\theta + 2\sin\theta\cos\theta$; $1/\cos\theta + 1/\sin\theta = b$
$a^2 = 1 + 2\sin\theta\cos\theta$; $\sin\theta + \cos\theta = b\sin\theta\cos\theta$
$a^2 - 1 = 2\sin\theta\cos\theta$; $b(a^2-1)/2 = a$; $b(a^2-1) = 2a$

Sol: 3
$7\text{cosec}\theta = 7 + 3\cot\theta$; $49\text{cosec}^2\theta = 49 + 9\cot^2\theta + 42\cot\theta$
$49(1+\cot^2\theta) = 49 + 9\cot^2\theta + 42\cot\theta$
$40\cot^2\theta = 42\cot\theta$
$\cot\theta = 21/20$; $\Rightarrow \text{cosec}\theta = 29/20$
Therefore, $(7 \times 21/20) - (3 \times 29/20) = 3$

Sol: 4
$a^2 - b^2 = (a+b)(a-b)$; $\Rightarrow a\sec\theta - b\tan\theta = 5$
Therefore, $a\sec\theta = 3$ and $b\tan\theta = -2$
$a^2\sec^2\theta = 9$; $a^2(1+\tan^2\theta) = 9$
$a^2(1+4/b^2) = 9$
$a^2b^2 + 4a^2 = 9b^2$

Sol: 5

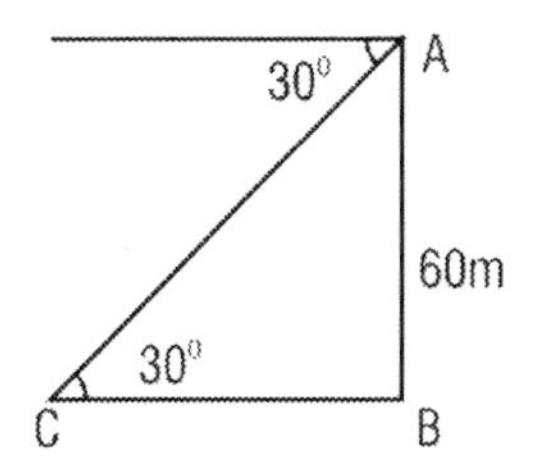

$\tan 30^\circ = 1/\sqrt{3} = 60/BC$; => $BC = 60\sqrt{3}$.

Sol: 6

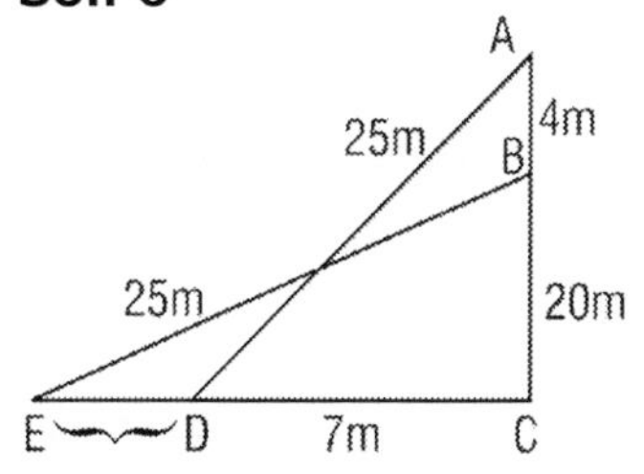

$AD = BE = 25$; $AC^2 = AD^2 - CD^2 = 625 - 49 = 576$
Therefore, $AC = 24$; => $BC = 20$
$BE^2 - BC^2 = AC^2$; $625 - 400 = 225$; => $BE = 15$ and $DE = 8$

Sol: 7

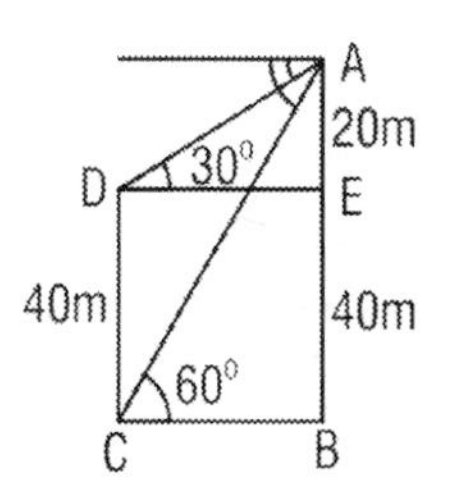

$\tan 60^\circ = \sqrt{3} = 60/BC$; => $BC = 20\sqrt{3} = DE$
$\tan 30^\circ = 1/\sqrt{3} = AE/20\sqrt{3}$; => $AE = 20$ and $BE = 40 = CD$

Sol: 8

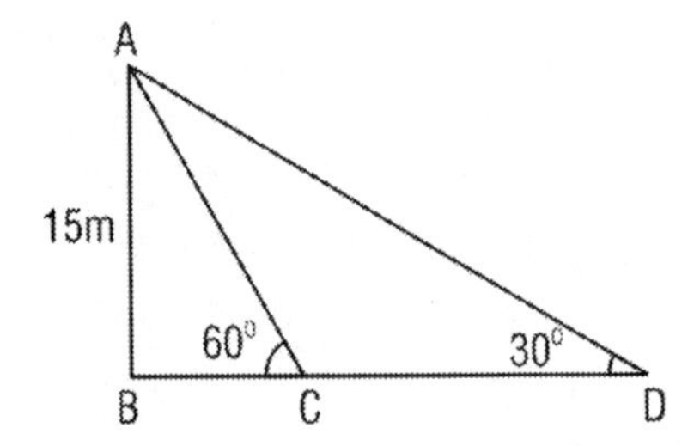

$\tan 30^\circ = 1/\sqrt{3} = 15/BD$; => $BD = 15\sqrt{3}$
$\tan 60^\circ = \sqrt{3} = 15/BC$; => $BC = 15/\sqrt{3}$
$BD - BC = 10\sqrt{3}$.

---

11. PROGRESSIONS

ARITHMETIC PROGRESSION

a, a+d, a+2d, a+3d,…………, a+(n−1)d are the terms of an AP, where a is the first term and d is the common difference. A_n is the nth term and it is given by a formula
$A_n = a + (n-1)d$

S_n is the sum of n terms of an AP and it is given by a formula
$S_n = \frac{n}{2}\{2a + (n-1)d\}$ or $\frac{n}{2}\{A_1 + A_n\}$

GEOMETRIC PROGRESSION

a, ar, ar^2, ar^3,………………., ar^{n-1} are terms of a GP, where a is the first term and r is the common ratio. The nth term is given by a formula
$A_n = ar^{n-1}$
S_n is the sum of n terms of a GP and it is given by a formula
$S_n = a(r^n-1)/(r-1)$, when $r>1$ or $a(1-r^n)/(1-r)$, when $r<1$
When $r<1$, the sum of infinite terms of a GP is given by a formula
$S_\infty = a/(1-r)$

HARMONIC PROGRESSION

If you take the reciprocal of the terms of harmonic progression it will form arithmetic progression, or the reciprocal of the terms of arithmetic progression will form harmonic progression. Therefore, you can apply the same formulae given in AP.

EG: 1, 1/2 and 1/3 are in harmonic progression because the reciprocals 1, 2 and 3 are the three terms of an AP.

SOLVED EXAMPLES

1. The interior angles of a polygon are in AP. The smallest angle is 120° and the common difference is 5. Then the number of sides of the polygon is:

Solution:

The interior angles are in AP: 120, 125, 130, 135 and so on
Let the number of sides of the polygon be n
Therefore, $n/2[2a+(n-1)d] = n/2[240+(n-1)5]$
We also know that sum of interior angles of any polygon is $(n-2)180$.
Equating both the above we get $n^2-25n+144 = 0$
$(n-9)(n-16) = 0$
As n cannot be 16, the value of n is 9.

2. A man arranges to pay off a debt of Rs.3600 in 40 instalments. The instalments form an arithmetic progression. When 30 of the instalments are paid, he dies leaving one-third of the debt unpaid. Find the value of the first instalment.

Solution:

$S_{30} = 15(2a+29d) = 2400$ or $2a+29d = 160$
$S_{40} = 20(2a+39d) = 3600$ or $2a+39d = 180$
Therefore, $d = 2$ and $a = 51$

3. What is the next number in the series given below?
53, 48, 50, 50, 47

Solution:

There are two sequences in the above series.
53, 50, 47, 44.......and so on
48, 50, 52, 54........and so on
The next number is the third number of the second sequence.

4. In a GP, the first term is 5 and the common ratio is 2. The eighth term is:

Solution:

@@The nth term of a GP is ar^{n-1}
Therefore, $5x2^7 = 5x128 = 640$

5. If the arithmetic mean of two numbers is 5 and geometric mean is 4, then the numbers will be:

Solution:

@@Arithmetic mean of two numbers 'a' and 'b' is (a+b)/2 and their geometric mean is $\sqrt{ab}$.

Therefore, a+b = 10; ab = 16

Apply the formula $(a+b)^2$ and $(a-b)^2$ to find a+b and a−b.

a= 8 and b = 2

6. Let S_n denote the sum of the first 'n' terms of an AP. If $S_{2n} = 3S_n$ then the ratio S_{3n}/S_n is equal to

Solution:

$S_n = n/2[2a+(n-1)d]$; $S_{2n} = 2n/2[2a+(2n-1)d] = 3n/2[2a+(n-1)d]$

Therefore, $2[2a+(2n-1)d] = 3[2a+(n-1)d]$

$4a+4nd-2d = 6a+3nd-3d$

$2a = nd+d$

$S_{3n} = 3n/2[2a+(3n-1)d]$; $S_n = n/2[2a+(n-1)d]$

Substituting the value of $2a = nd+d$

We get, $S_{3n}/S_n = 6$

7. The missing number in the series 8, 24, 12, 36, 18, 54,

Solution:

8x3 = 24÷2 = 12x3 = 36÷2 = 18x3 = 54÷2 = 27

8. The sum of the 6th and 15th elements of an AP is equal to the sum of 7th, 10th and 12th elements of the same progression. Which element of the series should necessarily be equal to zero?

Solution:

$T_6 = a+5d$; $T_{15} = a+14d$;

$a+5d+a+14d = a+6d+a+9d+a+11d$

$2a+19d = 3a+26d$

$a = -7d$

$T_8 = a+7d = 0$

EXERCISE

1. In a GP, the sum of the first and the last term is 66 and the product of the second and the last but one term is 128. Determine the first term of the series.

2. A sequence is generated by the rule that the nth term is n^2+1 for each positive integer n. In the sequence, for any value n>1, the value of $(n+1)^{th}$ term less the value of nth term is

3. What is the next term in the following series?
   2, 15, 41, 80

4. x, y and z are in geometric progression and the expression $x^{b-c} \times y^{c-a} \times z^{a-b}$ is equal to 1, then a, b and c are in
a) Geometric progression
b) Arithmetic progression
c) Harmonic progression
d) None of these

5. If x<1, the sum of the series $1+2x+3x^2+4x^3+$............to infinity is:

6. The sum of n terms of two arithmetic series are in the ratio of 3n+1:2n+1. What is the ratio of their 9th terms?

7. If x = $\sqrt{6+\sqrt{6+\sqrt{6+......\infty}}}$ what is the value of x?

8. If the sum of the n terms of an arithmetic progression is $3n^2+2n$, what is the rth term?

9. Three numbers are in GP. Their sum is 28 and their product is 512. Find the numbers.

10. If $\frac{1}{b-a}+\frac{1}{b-c}=\frac{1}{a}+\frac{1}{c}$, then a, b and c are in
a) Arithmetic progression
b) Geometric progression
c) Harmonic progression
d) None of these

SOLUTIONS

Sol: 1

$a+ar^{n-1}=66$; $ar\times ar^{n-2}=128$;=> $a^2r^{n-1}=128$

Therefore, $a+\frac{a\times 128}{a^2}=66$

$a+\frac{128}{a}=66$

Solving a = 64 or 2

Sol: 2

$(n+1)^{th}$ term = $(n+1)^2+1$ = n^2+2n+2

n^{th} term = n^2+1

Therefore the required value is 2n+1

Sol: 3

2+13 = 15+2x13 = 41+3x13 = 80+4x13 = 132

Sol: 4

y = $\sqrt{xz}$ = $(xz)^{1/2}$ (y is the geometric mean of x and z)

Substituting the value of y in the given expression, we get

$x^{b-c}\times(xz)^{\frac{c-a}{2}}\times z^{a-b}$

$x^{b-\frac{a+c}{2}}\times z^{\frac{a+c}{2}-b}$ = 1

$x^{b-\frac{a+c}{2}}=z^{b-\frac{a+c}{2}}$

This is possible only if $b-\frac{a+c}{2}=0$

Therefore, b = $\frac{a+c}{2}$ or $2b=a+c$

Hence a, b and c are in AP.

Sol: 5

The given expression can be written as:

$(1+x+x^2+x^3+.......)+(x+x^2+x^3+........)+(x^2+x^3+x^4+.....)+..........$

$\frac{1}{1-x}+\frac{x}{1-x}+\frac{x^2}{1-x}+.........=\frac{1+x+x^2+}{1-x}$ = $\frac{1}{(1-x)^2}$ ($\because$ in GP $S_\infty=\frac{a}{1-r}$, r<1)

Sol: 6

Let $S_1 = n/2[2a_1+(n-1)d_1]$ and $S_2 = n/2[2a_2+(n-1)d_2]$

For n = 17, we get

$S_1/S_2 = (a_1+8d_1)/(a_2+8d_2)$, which is the ratio of their 9th terms.

If we substitute n = 17 in 3n+1:2n+1, we get 52:35

Sol: 7

Since the sequence goes infinitely x will be equal to $\sqrt{6+x}$

Squaring both sides we get $x^2 = 6+x$ or $x^2-x-6 = 0$

$(x-3)(x+2) = 0$; => x = 3 or −2

x cannot be negative. Therefore, x = 3

Sol: 8

Sum of r terms = $3r^2+2r$

Sum of r−1 terms = $3(r-1)^2+2(r-1) = 3r^2-6r+3+2r-2 = 3r^2-4r+1$

$$Therefore, r^{th} term = sum of r terms – sum of (r-1) terms

6r-1

Sol: 9

Let the three terms are a/r, a, ar

$a^3 = 512$; => a = 8

(8/r)+8+8r = 28; => r = 2 or ½

Therefore, the numbers are 4, 8 and 16 or 16, 8 and 4

Sol: 10

$$\frac{1}{b-a}-\frac{1}{c}=\frac{1}{a}-\frac{1}{b-c}$$

Simplifying we get $b = \frac{2ac}{a+c}$, which implies a, b and c are in harmonic progression.

---

12. PERMUTATION AND COMBINATION

In general, permutation is used for arrangement and combination is used for selection.

For example, if you take two at a time from three things A, B and C and arrange it you will have 6 possibilities: AB, BA, AC, CA, BC and CA. Precisely, the order is important and it makes a difference. This is nothing but selecting two from three and then arrange it. You can select two things from three in 3 ways: AB, BC and CA. Each of the above can be arranged in 2 ways. Therefore, the total arrangements in the above example are 3x2 = 6 or technically, ${}^3P_2$ or ${}^3C_2$x2!

If the order is not important then you use combination. In the above example when you select two from three, the possibilities are AB, BC and CA. In certain cases, AB and BA will not make any difference. Here you have to use combination. The total number of selections will be ${}^3C_2$ ways or 3 ways.

Where the order is important: when you select two persons from three and seat them in two chairs, AB and BA will make difference.

Where the order is not important: when you select two persons from three for a committee, AB and BA will not make any difference.

The number of arrangements of r things from n distinct things is ${}^np_r$ = $\frac{n!}{(n-r)!}$

EG: From the word, 'NOT' how many two-letter words, meaningful or otherwise, can be formed without repetition.

Solution:
You have to find the number of arrangements of two letters from the word NOT which is given by the formula ${}^3P_2$ = 3x2 = 6.
(NO, NT, ON, OT, TN, TO)

Alternative method:
For a two-letter word, there are two places. The first place can be occupied by any of the three letters in 3 ways **and** the second place can be occupied by the remaining two letters in 2 ways. Therefore, 3x2 = 6 ways (Remember, 'and' means multiplication)

The number of arrangements of r things from n distinct things when repetition is allowed is n^r

EG: In the above example, how many two-letter words can be formed with repetition?
Solution:
In addition to the 6 words you also have NN, OO and TT. Altogether, you have 9 arrangements. This is given by the formula $3^2 = 9$.
Alternative method:
The first place can be occupied by any of the three letters in 3 ways **and** the second place is occupied by any of the three letters in 3 ways. Hence, 3x3 = 9 ways

The number of arrangements of n things, taking all at a time, when one thing is repeated p times, another repeated q times and another repeated r times = $\frac{n!}{p!q!r!}$.

EG: How many three-letter words, meaningful or otherwise, can be formed using the letters of the word TOO?
Solution:
In the word TOO, O is repeated two times. Hence, using the formula we get 3!/2! = 3.
(TOO, OTO, OOT)

The number of selections of r things from n distinct things is ^nc_r = $\frac{n!}{r!(n-r)!}$

n! = n(n−1)(n−2)(n−3)……3.2.1
Remember 0! = 1 and 1! = 1

EG: There are 3 tennis-players. You have to form a team of two to play a tournament. In how many ways can it be done?

Solution:
Let the three players be A, B and C. There are 3 ways of making the team with two players, AB, BC and CA. This can be done using the formula $^3C_2 = 3$.

EG: There are 3 boys and 2 girls. A committee of 2 boys and 1 girl is to be formed. In how many ways can it be done?
Solution:
Two boys can be selected from the three boys in $^3C_2$ ways **and** 1 girl can be selected from the two girls in $^2C_1$ ways. Therefore, the total number of selections is $^3C_2 x ^2C_1 = 6$

EG: In the above example, if a committee of 3 persons is to be formed then in how many ways can it be done?
Solution:
The three persons can be all 3 boys **or** 2 boys and 1 girl **or** 1 boy and 2 girls.
Therefore, you have three possibilities.
Therefore, $^3C_3 + (^3C_2 x ^2C_1) + (^3C_1 x ^2C_2)$ ways it can be done.
1+6+3 = 10 ways
(Remember **and** means multiplication and **or** means addition)

The total number of selections from n distinct things: $^nC_0 + ^nC_1 + ^nC_2 + \ldots\ldots\ldots + ^nC_n = 2^n$
Remember, in the above $^nC_0$ indicates selecting none from n things and it is equal to 1. **Selecting one or more things from n distinct things is equal to $2^n - 1$.**
$^nC_0 = ^nC_n = 1$
$^nC_1 = n$; $^nP_1 = n$
$^nC_2 = n(n-1)/2!$; $^nP_2 = n(n-1)$
$^nC_3 = n(n-1)(n-2)/3!$; $^nP_3 = n(n-1)(n-2)$

EG: In an examination, there are 10 questions. In how many ways a student can answer these questions.
Solution:
A student may answer none of the questions in 1 way ($^{10}C_0$)
A student may answer only one question in 10 ways ($^{10}C_1$)
A student may answer two questions in 45 ways ($^{10}C_2$) and so on.
You can use the formula: 2^{10}.
In the above example the student can answer one or more questions in $2^{10} - 1$ ways.

EG: In an objective type test, there are 10 questions. Each question has two choices, either yes or no. All questions are compulsory. In how many ways a student can answer.

Solution:

Each question can be answered in two ways. Therefore, the answer is 2^{10}

The total number of ways in which a selection can be made by taking some or all out of p+q+r+.... things where p are all of one kind, q are all of a second kind, r are all of a third kind and so on is {(p+1)(q+1)(r+1).....} – 1

The number of ways 'n' things can be arranged around a circular table = (n−1)!

Note: If you arrange persons in a circular table, the clockwise and anti-clockwise arrangement makes a difference. In certain cases, clockwise and anti-clockwise arrangement will not make a difference: arrangement of beads in a necklace. In such cases, 'n' things can be arranged around a circle in $\frac{(n-1)!}{2}$ ways.

SOLVED EXAMPLES

1. If two dice are tossed simultaneously, the number of elements in the resulting sample space is:
Solution:
There are 6 numbers written in a dice from 1 to 6. The number of all possible outcomes is 6^1 in case of a single dice, 6^2 in case of two dice, 6^3 in case of three dice and 6^n in case of n dice.

2. A boy has 3 library cards and 8 books of his interest in the library. Of these 8, he does not want to borrow chemistry part 2 unless chemistry part 1 is also borrowed. In how many ways can he choose the three books to be borrowed?
Solution:
There are two cases.
He selects chemistry part 2:
It implies that he will also borrow chemistry part 1. The remaining one book can be selected from the remaining 6 books in $6C_1$ ways or 6 ways.
He does not select chemistry part 2:
In this case, he can select the three books from the remaining 7 books in $7C_3$ ways or 35 ways.
Therefore, the total number of selections is 6+35 = 41

3. Four different objects 1,2,3,4 are distributed at random in four places marked 1,2,3,4. In how many ways none of the objects occupy the place corresponding to its number.
Look at the following table:

1	2	3	4
2	1	4	3
	3	4	1
	4	1	3
	4	3	1

When 2 occupy the first place, there are 4 ways that all the objects are in different places. You get the same possibility when 3 and 4 occupy the first place. Therefore 3x4= 12, is the number of ways that none of the objects occupy the place corresponding to its number.

4. There are 6 positive and 8 negative numbers. Four numbers are chosen at random and multiplied. In how many ways the resulting number is positive?

Solution:

There are three cases:

All four are positive numbers. 4 positive numbers can be selected in $6C_4$ ways or 15 ways.

All four are negative numbers. 4 negative numbers can be selected in $8C_4$ ways or 70 ways.

Two are positive and two are negative numbers. They can be selected in $6C_2 x 8C_2$ ways or 420 ways.

Hence the total number of ways is 15+70+420 = 505.

5 .Two dice are tossed. If the scores are added, in how many cases you get a prime number.

Solution:

The minimum and maximum scores are 2 and 12 respectively.

2,3,5,7 and 11 are prime numbers.

2 -> (1, 1)

3 -> (1, 2), (2, 1)

5 -> (1, 4), (4, 1), (2, 3), (3, 2)

7 -> (1, 6), (6, 1), (2, 5), (5, 2), (3, 4), (4, 3)

11-> (5, 6), (6, 5)

You get a prime number in 15 cases.

6. Ten different letters of an alphabet are given. Words with 5 letters are formed from these given letters. Then the number of words, which have at least one letter repeated, is:

Solution:

(Total number of words that can be formed) – (total number of words with no letter is repeated) will be the required answer.

The five letter word has five places. Each place can be occupied by any of the 10 letters in 10x10x10x10x10 = 10^5 ways

The total number of ways with no letter is repeated is 10x9x8x7x6 ways or $10P_5$

The required answer is 10000–30240 = 69760.

7. Eight chairs are numbered from 1 to 8. Two women and three men wish to occupy one chair each. First, the women chose the chairs from amongst the chairs marked 1 to 4. Then the men selected the chairs from amongst the remaining marked from 5 to 8. The number of possible arrangements is:

Solution:
The number of arrangements for the two women to occupy chairs from 1 to 4 is $4P_2$ ways.
The number of arrangements for the three men to occupy chairs from 5 to 8 is $4P_3$ ways.
Therefore, the total number of arrangements is $4P_2 x 4P_3$ ways.

8. In a hockey championship there were 153 matches played. Every two teams played one match with each other. The number of teams participating in the championship is:

Solution:
PROBLEM SOLVING TECHNIQUE
We can generate a formula for this type of problems.
If there are two teams, 1 match will be played; 1
If there are 3 teams, 3 matches will be played; 2+1
If there are 4 teams, 6 matches will be played; 3+2+1
If there are n teams, (n−1) + (n−2) + (n−3) +......+3+2+1 matches will be played.
This is nothing but sum of first (n−1) natural numbers.
The formula is n (n−1)/2.
Therefore, n (n−1) = 306 = 18x17; => n = 18
#Note that the above formula can also be used for the number of handshakes.

9. In an examination paper, there are two groups each containing four question. A candidate is required to attempt 5 questions but not more than 3 questions from any group. In how many ways can 5 questions be selected?

Solution:
Either he can chose 3 from first group and 2 from the second or 2 from the first and 3 from the second.
In the first case it can be done in $4C_3 x 4C_2$ ways
In the second case it can be done in $4C_2 x 4C_3$ ways.
Therefore, (4x6) + (6x4) = 48 ways.

10. A box contains 10 distinct balls out of which 3 are red and the rest are blue. In how many ways can a random sample of 6 balls be drawn from the bag so that at the most 2 red balls are included in the sample and no sample has all the 6 balls of the same colour?

Solution:
There are two cases. It can be either 1 red and 5 blue or 2 red and 4 blue.
The first case can be done in $3C_1x7C_5$ ways.
The second case can be done in $3C_2x7C_4$ ways.
Therefore, 3x21 + 3x35 = 168 ways

EXERCISE

1. Three boys and three girls are to be seated around a table in a circle. Among them the boy X does not want any girl neighbour and the girl Y does not want any boy neighbour. How many such arrangements are possible?

2. Two series of a question booklet for an aptitude test are to be given to twelve students. In how many ways can the students be placed in two rows of six each so that there should be no identical series side by side and that the students sitting one behind the other should have the same series?

3. A class photograph has to be taken. The front row consists of 6 girls who are sitting. 20 boys are standing behind. The two corner positions are reserved for the 2 tallest boys. In how many ways can the students be arranged?

4. An intelligence agency decides on a certain code of 2 digits selected from 0 to 9. However, the slip on which the code is hand-written allows confusion between top and bottom, because these are indistinguishable. Thus, for example, the code 91 could be confused with 16. How many codes are there such that there is no possibility of any confusion?

5. A five-digit number is formed using digits 1, 3, 5, 7 and 9 without repeating any one of them. What is the sum of the all such possible numbers?

6. Consider the five points comprising the vertices of a square and the intersection point of its diagonals. How many triangles can be formed using these points?

7. How many different 5 digit numbers can you make using 1, 2, 3, 4 and 5 (without repetition) such that the digit in the unit's place is always greater that the digit in the hundred's place?

a) 60 b) 45 c) 30 d) None of these

8. Ten points are marked on a straight line and eleven points are marked on another straight line. How many triangles can be constructed with vertices from among the above points?
a) 495 b) 550 c) 1045 d) 2475

9. One red flag, three white flags and to blue flags are arranged in a line such that,
a) No two adjacent flags are of the same colour
b) The flags at the two ends of the line are of different colours.
In how many different ways can the flags be arranged? (CAT, 2000)
a) 6 b) 4 c) 10 d) 2

10. Sam has forgotten his friend's seven digit telephone number. He remembers the following: the first three digits are either 635 or 674; the number is odd; and the number 9 appears only once. If Sam reaches his friend, what is the minimum number of trials he has to make before he can be certain to succeed? (CAT, 2000)
a) 10000 b) 2430 c) 3402 d) 3006

SOLUTIONS

Sol: 1

Put the boy X in the middle of the other boys and the girl Y in the middle of the other girls. The seats of X and Y are fixed in one way. The remaining two boys can be seated in two ways and the remaining girls can be seated in two ways. Therefore, this arrangement is possible in 2x2 = 4 ways.

Sol: 2

Let us arrange the students first.
There are two rows 1 and 2. Six students are to be arranged in first row and six in the second row. Select 6 from the 12 and arrange them in the first row. Six students can be selected in $12C_6$ ways and they can be seated in 6! ways. The remaining six students can be seated in 6! ways. Therefore the total arrangements is $12C_6$x6!x6!
Now let us arrange the booklets.
There are two series A and B. You can start with A and this is one arrangement. You can also start with B in another arrangement. There are only two such possibilities.
Hence the required number of arrangements is 2x$12C_6$x6!x6!

Sol: 3

The six girls can be seated in 6! ways. The two tallest boys can be arranged in 2 ways.
The remaining 18 boys can be arranged in 18! ways. Therefore the total arrangements = 1440x18!

Sol: 4

The total number of two-digit numbers using digits from 0 to 9 is $10\times10 = 100$
The numbers, which are confusing, are 0, 1, 6, 8 and 9
From these 5 digits we can form 25 two-digit numbers.
Out of these 00, 11 and 88 are not confusing.
Therefore, there are 22 numbers, which are confusing. Hence, the remaining 78 are not confusing.

Sol: 5

As repetition is not permitted, the total number of all such possible numbers is 5! It is equal to 120.

In the unit place of all such numbers there will be twenty four 1's, twenty four 3's and so on. The sum will be equal to 24(1+3+5+7+9) = 600. When you add the unit places of all numbers the resulting unit place is 0 and you carry 60. The sum of ten's digits plus the carry is 660. Therefore, the ten's digit of the sum is 0 and the carry is 66. The sum of hundred's digits and carry is 666. Therefore, the hundred's digit is 6 and carry is 66 and so on. Finally you get 6666600 as the answer.

Sol: 6

There are five points. To draw a triangle you need 3 points. Therefore, select 3 points from the five. This can be done in $5C_3$ ways or 10 ways. Out of these, you cannot select the two diagonals in which all the three points are collinear. Therefore, 8 triangles can be formed.

Sol: 7

'1' cannot occupy the unit's place. When '2' occupies the unit's place only '1' can occupy the hundred's place. The remaining 3 numbers in three places can be arranged in 3! ways. Look at the following table:

Unit's place	Hundred's place	No. of arrangements
1	Nil	Nil
2	1	6x3!
3	1, 2	12(3! + 3!)
4	1, 2, 3	18(3! + 3! + 3!)
5	1, 2, 3, 4	24(4x3!)

Option 'a' is the correct answer.

Sol: 8

To draw a triangle you need three points. Select two points from the first line and one point from the second line. This can be done in $10C_2 \times 11C_1$ ways. In addition, you can select two points from the second line and one point from the first line. This can be done in $11C_2 \times 10C_1$ ways. Therefore, the total no. of triangles will be 1045.

Sol: 9

Look at the following arrangements:

W R W B W B
W B W B W R
W B W R W B

These three arrangements can be reversed to get another three arrangements. Hence, the correct answer is 6.

Sol: 10

The first three places are occupied by either 635 or 674. Therefore, we have to fill the remaining four places. Since the number is odd, the last digit has to be odd. There are five odd numbers. It is also given that the number 9 appears only once. If 9 is the last number, then the remaining three places can be occupied by any of the numbers from 0 to 8. This can be done in 9^3 ways or 729 ways.

If 9 is not the last number, the place can be occupied by 1 or 3 or 5 or 7. Out of the remaining three places 9 will occupy one place in 3 ways. The remaining two places can be occupied by any of the numbers from 0 to 8 in 9^2 ways or 81 ways. Therefore, the above arrangement can be done in $4 \times 3 \times 81$ or 972 ways. Altogether there are 1701 arrangements if the first three digits are 635. If the first three digits are 674, there are another 1701 arrangements. Hence, the answer is 3402.

13. PROBABILITY

In an experiment, if the result is not unique and may be one of several possibilities then you say that the result is probable. Probability deals only with such experiments where the outcome is not unique.
When an experiment is conducted under identical conditions and if the result is one of several possible outcomes then the experiment is called a **random experiment**.
Each of the outcomes is called an **event**.

EG: Tossing of an unbiased coin.
Tossing of an unbiased coin is a random experiment, because the result is not certain. It can be head or tail. Getting a head is an event. Similarly, getting a tail is an event.
There are only two events in this random experiment. They are together called the **sample space** of the experiment.
In the above example the chance of getting a head and getting a tail are equal. Therefore, they are called **equally likely** events.

Difference between occurrence and event
All the possible occurrences of an experiment are events. However, an event may be defined to contain one or more occurrences.
EG: When a dice is cast there are 6 occurrences. Each occurrence is called an event. In addition to that an event may be defined to be the happening of an odd number. This event contains three occurrences.

Events that together cover all the possible occurrences of an experiment are termed as **collectively exhaustive.**
In the above example, if E_1 is defined as the happening of an odd number and E_2 is defined as the happening of an even number then E_1 and E_2 together are collectively exhaustive.
$E_1 = \{1, 3, 5\}$; $E_2 = \{2, 4, 6\}$
In the above example the occurrences of E_1 are not appearing in the event E_2 and vice versa. Hence, the two events are called **mutually exclusive.**
Therefore, the events E_1 and E_2 are **mutually exclusive and collectively exhaustive.**
Note that two events, which are mutually exclusive, are not necessarily collectively exhaustive.

Suppose you have two balls, one is red and the other is green, in a bag. You draw the two balls one by one. Let the first draw be E_1 and the second draw be E_2. Here, there are two cases: **with replacement and without replacement**.

With replacement means after the first draw, you replace the ball drawn in the bag and then you draw the second ball.

Without replacement means after the first draw you keep aside the ball drawn and then draw the second ball.

In the first case the event E_2 does not depend upon the event E_1. E_1 and E_2 are **independent events.**

In the second case the event E_2 depends upon the event E_1. If the first ball is red the second ball will be green and vice versa.

Probability defined

If in an experiment the total number of occurrences is 'n' and an event(E) is defined to contain 'm' occurrences then the probability of E or P(E) = m/n.

$P(\overline{E})$ = non occurrence of the event E = $\frac{n-m}{n}$ or $1 - \frac{m}{n}$.

EG: when a die is thrown what is the probability that the number is odd?

Solution:

The number of occurrences that are favourable to the event is 3(1, 3 and 5)

The total number of occurrences is 6.

Therefore, P(E=odd) = 3/6 = ½.

In the above example, the probability that the number is not odd is $1-\frac{1}{2}=\frac{1}{2}$.

Note that in an experiment if an event will never happen then its probability is 0 and if an event is certain to happen then its probability is 1. Therefore, probability of any event will be greater than or equal to 0 and less than or equal to 1.

$0 \le P(E) \le 1$

Addition theorem

If A and B are two events which are not mutually exclusive then probability of A or B or $P(A \cup B) = P(A)+P(B)-P(A \cap B)$.

If A and B are two events which are mutually exclusive then $P(A \cup B)=P(A)+P(B)$.

EG: If two cards are drawn simultaneously from a pack of cards what is the probability that both are spades or clubs?

Solution:

Here the two events are mutually exclusive.

P(E1=spades)= $^{13}C_2/^{52}C_2$

P(E2=clubs)=$^{13}C_2/^{52}C_2$

P(E1 or E2)=P(E1$\cup$E2)=$\frac{^{13}C_2}{^{52}C_2}+\frac{^{13}C_2}{^{52}C_2}$

EG: If two cards are drawn simultaneously from a pack of cards what is the probability that both are aces or both are red?

Solution:

Here, the two events are not mutually exclusive because out of four aces in a pack of cards two are red. The two red aces are common in both the events. Remember, in a pack of cards there are 26 red cards and 26 black cards.

Therefore, the required probability is $\frac{^{4}C_2}{^{52}C_2}+\frac{^{26}C_2}{^{52}C_2}-\frac{^{2}C_2}{^{52}C_2}$

Independent events

If the occurrence of one event is not affected by the occurrence or non-occurrence of another event then they are independent events. If A and B are two independent events then P(A$\cap$B)=P(A).P(B)

EG: If two cards are drawn one by one from a pack of cards **with replacement**, what is the probability that both are aces?

Solution:

Let E1 be the event that the first card drawn and E2 be the event that the second card drawn.

You have to find the probability of E1=ace **and** E2=ace or P(E1$\cap$E2).

Therefore, the required probability is $\frac{^{4}C_1}{^{52}C_1}\times\frac{^{4}C_1}{^{52}C_1}$

Conditional probability

When two events are defined in an experiment, the probability of one event given that the other event has already happened is called conditional probability.

Let A and B are two events. The probability of B given that the event A has already happened is P(B)= $\frac{P(A\cap B)}{P(A)}$

SOLVED EXAMPLES

1. A coin is tossed 5 times. What is the probability that head appears an odd number of times?

Solution:

The sample space is:

{(5H,0T),(4H,1T),(3H,2T),(2H,3T),(1H,4T),(0H,5T)}

Out of these the favourable cases are 3.

Therefore, the probability is ½.

2. Atul can hit a target 3 times in 6 shots, Bhola can hit the target 2 times in 6 shots and Chandra can hit the target 4 times in 4 shots. What is the probability that at least two of them hit the target?

Solution:

As Chandra hits the target definitely, either one of Atul and Bhola or both hit the target.

The probability of Atul hitting the target and Bhola not hitting the target is 3/6 x 4/6

The probability of Atul not hitting and Bhoal hitting the target is 3/6 x 2/6

The probability of both hitting the target is 3/6 x 2/6

Therefore, the required probability is 1/3 + 1/6 + 1/6 = 2/3

3. A bag contains 5 white, 7 red, and 8 black balls. If 4 balls are drawn one by one with replacement, what is the probability that all are white?

Solution:

The probability of the first ball being white is $5C_1/20C_1$ = ¼.

Since the ball is replaced each time the probability will remain the same for the next three draws. Therefore, the required probability is ¼ x ¼ x ¼ x ¼ = 1/256.

4. A dice is thrown 6 times. If getting an odd number is a success then the probability of 5 successes is:

Solution:

Let S be the success and F be the failure.

SSSSSF is the required case. However, failure can happen at any one of the throws. Therefore, there are 6 such cases favourable. The probability of S is ½ and the probability F is ½ as there are equal number of odd numbers and even numbers.

Hence the required probability is 6 x $(½)^6$ = 3/32.

5. A bag has 4 red and 5 black balls. A second bag has 3 red and 7 black balls. One ball is drawn from the first bag and two from the second. The probability that there are two black balls and a red ball is:

Solution:

The one ball, which is drawn from the first bag, can be either a red or black.

If the ball is red then you require 2 black balls from the second bag.

If the ball is black then you require 1 black and 1 red from the second.

Therefore, the required probability is $(4C_1/9C_1)x(7C_2/10C_2) + (5C_1/9C_1)x(3C_1x7C_1/10C_2) = 7/15$

EXERCISE

1. From amongst 36 teachers in a school, one principal and one vice-principal are to be appointed. In how many ways can this be done?

2. A bag contains 3 white balls and 2 black balls. Another bag contains 2 white balls and 4 black balls. A bag and a ball are picked at random. The probability that the ball will be white is:

3. One hundred identical coins each with probability p showing up heads are tossed. If $0<p<1$ and the probability of heads showing on 50 coins is equal to that of heads on 51 coins, then the value of p is:

4. The probability that a student is not a swimmer is 1/5. Then the probability that out of the five students, four are swimmers is:

5. A bag contains 2 red, 3 green and 2 blue balls. 2 balls are to be drawn randomly. What is the probability that the balls drawn contain no blue ball?

SOLUTIONS

Sol: 1

2 persons from 36 can be selected in $36C_2$ ways.
The selected 2 can be appointed in 2 ways.
Therefore, in $2x36C_2$ ways or 1260 ways it can be done.

Sol: 2

The probability of the first bag being picked up is ½
The probability of the ball picked up being white is 3C1/5C1
The probability of the second being picked up is ½
The probability of the ball picked up being white is 2C1/6C1
Therefore, the required probability is ½(3/5 +2/6) = 7/15

Sol: 3

$$When the number of trials is more, we generally use binomial distribution.
$nC_r.p^r.q^{n-r}$, where n is the number of trials, p is probability of success, and q is the probability of failure.
Here, the number of trials is 100. In the first case r = 50 and in the second case r = 51.
Therefore, $100C_{50}.p^{50}.q^{50} = 100C_{51}.p^{51}.q^{49}$

$$\frac{100!}{50!\times 50!}p^{50}\times(1-p)^{50} = \frac{100!}{51!\times 49!}p^{51}\times(1-p)^{49}$$

$$\frac{p}{1-p}=\frac{51!\times 49!}{50!\times 50!}=\frac{51\times 50!\times 49!}{50!\times 50\times 49!}=\frac{51}{50}$$

It implies p = 51/101

Sol: 4

Let S denotes the swimmer and F denotes the student who does not swim.
The favourable case is SSSSF.
The probability of S is 4/5 and F is 1/5.
In addition, F can occur at any of the five places.
Therefore, the required answer is $5x(4/5)^4x1/5$

Sol: 5

The two balls drawn can be 1 red and 1 green or 2 red or 2 green.

$$\frac{2c_1\times 3c_1}{7c_2}+\frac{2c_2}{7c_2}+\frac{3c_2}{7c_2}$$

2/7 + 1/21 + 1/7 = 10/21

14. INEQUALITIES, EXPONENTS, DESCRIPTIVE STATISTICS, AND FUNCTIONS

Any quantity 'a' is said to be greater than another quantity 'b' when a-b is positive. For example, 3 is greater than 2, because 3-2=1, which is positive. Also, 'b' is said to be less than 'a' when b-a is negative. For example, 2 is less than 3, because 2-3=-1, which is negative. Zero is greater than any negative quantity.

If $a>b$, then
$a+c > b+c$;
$a-c > b-c$;
$ac > bc$;
$a/c > b/c$;
that is, an inequality will still hold after each side has been increased, diminished, multiplied, or divided by the same positive quantity.

If $a-c>b$, then $a>b+c$ (by adding c to each side). It means, in an inequality any term may be transposed from one side to the other if its sign is changed.

If $a>b$, then evidently $b < a$; that is if the sides of an inequality are transposed, the sign of inequality must be reversed. E.g. $3>2$; $2<3$

If $a>b$, then $-a < -b$.
Proof:
$a-b>0$, and $b-a<0$;
Therefore, $-a-(-b)<0$ $(b=-(-b))$;
that is $-a<-b$ ($-(-b)$ is transposed).
Hence, if the signs of all the terms of an inequality are changed, the sign of inequality must be reversed. That is, if the sides of an inequality are multiplied by the same negative quantity, the sign of inequality must be reversed.
Thus if $a > b$, then $-ac < -bc$.
e.g. $3>2$; $-3<-2$ (each side is multiplied by -1)

If $a>b$, and if n is any positive quantity, then $a^n>b^n$.
Further, $a^{-n}<b^{-n}$.

e.g. if a=3, b=2, and n=2 then 9>4

if a=3, b=2, and n=-2 then $\frac{1}{9} < \frac{1}{4}$

The square of every real quantity is positive and therefore greater than zero. Thus $(a-b)^2$ is positive.
Therefore, $a^2-2ab+b^2>0$
$a^2+b^2>2ab$
In addition, $\frac{a+b}{2} > \sqrt{ab}$
That is the arithmetic mean of two positive quantities is greater than their geometric mean. The inequality becomes equality when quantities are equal.

If a, b, c denote positive quantities then $a^2+b^2+c^2>bc+ca+ab$

If the sum of two positive quantities is given, their product is greatest when they are equal; and if the product of two positive quantities is given, their sum is least when they are equal.
e.g. if the sum of two positive quantities is 10, then 5x5=25 is the greatest though 10 can be written as 1+9, 2+8, 3+7, 4+6; each of the product 1x9, 2x8, 3x7, 4x6 is less than 25.
Similarly, if 16 is the product of two positive quantities, then 4+4=8, 2+8=10, 1+16=17.

If a and b are positive and unequal, then
$\frac{a^m+b^m}{2} > \left(\frac{a+b}{2}\right)^m$; except when m is a positive proper fraction. If m=0 or 1, the inequality becomes an equality.

e.g. if a=1, b=2, and m=2 then 2.5>2.25
if a=4, b=9, and m=1/2 then 2.5<2.535
if a=1, b=4, and m=1 then 2.5=2.5

Examples

1.Solve 3x-4>4x-3

Solution:
3x-4x>-3+4
-x>1; x<-1 (multiplying each side with -1)

2.If $6x-5 \leq 7x-3$ and $7x-4 \leq 5x-1$ then find the common solution.

Solution:
$-x \leq 2$ and $2x \leq 3$
$x \geq -2$ and $x \leq 3/2$
$\therefore x \in [-2,3/2]$ (square bracket indicates that the values are inclusive)

3.Solve for real x: $2x^2+3x-9 \leq 0$

Solution:
$2x^2+6x-3x-9 \leq 0$
$2x(x+3)-3(x+3) \leq 0$
$(2x-3)(x+3) \leq 0$
$\therefore x \in [-3,3/2]$

4.If $(x+2)(x+4)(x+3)^2>0$, then find the solution set

Solution:
$(x+3)^2$ is positive since square of every real quantity is positive.
Therefore, $(x+2)(x+4)>0$
Hence, $x<-4$ or $x>-2$

5.Find the value of x in the inequality $x^2-6x+11>0$.

Solution:
The given inequality can be written as $(x-3)^2-9+11>0$
$(x-3)^2+2>0$
For any real value of x, the above inequality is true (square of any real quantity is positive; positive plus positive is positive).
$\therefore x \in R$

6.Solve for $x: \frac{1}{3x+4} \leq 0$

Solution:
$\frac{3x+4}{(3x+4)^2} \leq 0$ (multiplying numerator and denominator by 3x+4; now the denominator is positive and therefore the numerator is negative)

$\Rightarrow 3x+4 \le 0$

$3x \le -4$

$x \le -4/3$

$\therefore x \in \left(-\infty, -4/3\right]$

7.Solve for $x: \dfrac{2x^2+5x-12}{x^2+6x-16} < 0$

Solution:

$\dfrac{(2x-3)(x+4)}{(x+8)(x-2)} < 0$

$(2x-3)(x+4)(x+8)(x-2) < 0$ (multiplying numerator and denominator by (x+8)(x-2); now the denominator is positive and the numerator is negative)

Now, if you mark the possible intervals on a number line you get (-8, -4), (-4, 3/2), and (3/2, 2). The given expression is negative only in (-8, -4) and (3/2, 2).

$\therefore x \in (-8, -4) \cup (3/2, 2)$

After resolving the numerator and denominator, it is better to substitute the values in the expression from the options to find out the intervals.

Absolute Value and Properties of Modulus

For any real number x, the absolute value is defined as

$|x| = \begin{cases} x \text{ if } x \ge 0 \text{ and} \\ -x \text{ if } x < 0 \end{cases}$

Absolute value of x is written as $|x|$ and read as 'modulus x'.

For any real number x and y,

1.x=0 <=> $|x|=0$

2. $|x+y| \le |x| + |y|$

3. $|x-y| \ge |x| - |y|$

4. $|x.y| = |x|\,|y|$

Examples

1.If $|x+5|<9$, find the solution set for x.

Solution:
It implies $(x+5)<9$ or $-(x+5)<9$ (definition of absolute value)
Therefore, $x<4$ or $-x<14$
$x<4$ or $x>-14$
$x<4$ or $-14<x$ (if $x>y$, then $y<x$)
We can combine the above and write
$-14<x<4$
$\therefore x \in (-14, 4)$

2.Find the minimum value of the function $f(x)=5+|x+3|$

Solution:
For any real value of x, $|x+3|$ is always positive
Therefore, the function attains minimum when $|x+3|$ is zero
It implies $x=-3$
Hence the minimum value of the function is 5 at $x=-3$.

3.If $\left|\frac{4x-3}{2}\right| \leq 3$, find the solution set for x.

Solution:
$|4x-3| \leq 6$ (since the denominator is a constant and positive)
$4x-3 \leq 6$ or $-(4x-3) \leq 6$
$4x \leq 9$ or $-4x \leq 3$
$x \leq 9/4$ or $x \geq -3/4$
$-3/4 \leq x \leq 9/4$
Therefore, $x \in [-3/4, 9/4]$

4.If $\left|\frac{2}{4x-9}\right| \geq \frac{3}{7}$, then find the solution set for x.

Solution:
$\frac{2}{4x-9} \geq \frac{3}{7}$ or $\frac{2}{4x-9} \geq -\frac{3}{7}$
$14 \geq 12x-27$ or $14 \geq -12x+27$
$12x \leq 41$ or $12x \geq 13$
$x \leq 41/12$ or $x \geq 13/12$

In addition, the function is not defined at $x=9/4$

$\therefore x \in [13/12, 9/4) \cup (9/4, 41/12]$

5.Find the solution set of x for the real function $\sqrt{\frac{4-x}{2x-5}}$

Solution:

Multiply the numerator and denominator by (2x-5). The denominator is now positive. The numerator (4-x)(2x-5) cannot be negative.

Therefore, $(4-x)(2x-5) \geq 0$

$4-x \geq 0$ or $2x-5 \geq 0$

$-x \geq -4$ or $2x \geq 5$

$x \leq 4$ or $x \geq 5/2$

In addition the function is not defined at $x=5/2$

$\therefore 5/2 < x \leq 4$

Exponents

Suppose, if we add 1 horse+1 horse+ 1 horse we get 3 horses. Similarly, if we add 1 egg+1 egg+1 egg we get 3 eggs. The above is true for horse or egg or whatever thing we consider. Therefore, the rule can be generalized. Instead of saying 1 horse or 1 egg or 1 thing, we say 'x'. That is, 1x + 1x +1x is 3x. Mathematical language does not strictly follow grammar. We do not give a plural form of 'x'. It is understood. In mathematics, 'x' is called a variable which can assume anything say, a horse or an egg or even a number.

In the above example, '1x' can simply be written as 'x', and therefore, x + x + x = 3x. Suppose if we add one more x, then we get 4x, and if we add n number of times x then we get nx. Now, let us assume 'x' is 2 (a number). Then 2+2+2 is adding 3 times 2, which gives 6. In mathematics, we give a notation for '3 times 2' as '3 multiplied by 2' or simply we write 3x2. In general, if we add n number of times 2 we get nx2 or 2n (conventionally, we write 2n instead of n2). Now confusion occurs as to whether we should read as '2 times n' or 'n times 2'. In fact, both are different, and each means different thing. If we assume n as an egg then 2 times egg means adding 1 egg to 1 egg. However, 'egg times 2' makes no sense. Therefore, 'n' has to be defined. If 'n' is a number, say '3' then '3 times 2' makes sense. Therefore, when 'n' is a number, both 'n times 2' and '2 times n' give the same result. That is 3x2=2x3.

In the above example, if we replace the addition operator with multiplication operator then it leads us to a different notation. If we multiply three times 2 then 2x2x2 is 8 because we know '2 times 2' is 2x2=4; '2 times 2 times 2' is '4 times 2' which is 2+2+2+2=8. Hence, 2x2x2=2+2+2+2. This relationship between addition and multiplication is a great invention in the field of mathematics. In mathematics, multiplying three times 2 is written as 2^3 and multiplying 'n times 2' is 2^n. Remember we can multiply only numbers. We cannot do an egg x an egg to give egg^2; it makes no sense. When we multiply 'n times 2', 'n' is the exponent of the number 2 and it is written as 2^n. In general, multiplying 'n times x' is written as x^n.

1.Consider 'multiplying 3 times 2' multiplied by 'multiplying 4 times 2'. It is written as $2^3 x 2^4$. This is nothing but 2x2x2x2x2x2x2, which is equal to 7 times multiplying the number 2 or 2^7. Therefore, $2^3 x 2^4 = 2^7$.

If you observe it carefully, you can generalize this result: $2^m x 2^n = 2^{m+n}$. If you replace 2 with 'a', then you get $a^m x a^n = a^{m+n}$.

2.Similarly, $2^4 \div 2^3 = 2^1$. In the numerator, you have 2x2x2x2 and in the denominator, you have 2x2x2. When you cancel out the common factors, you get 2 as the answer. Therefore, $2^4 \div 2^3 = 2^{4-3} = 2^1 = 2$. In general, $a^m \div a^n = a^{m-n}$.

3.Consider $(2^4)^3$. It means 'multiplying three times 2^4', that is, $2^4 x 2^4 x 2^4$, which is equal to $2^{4+4+4} = 2^{3x4} = 2^{4x3}$. In general, $(a^m)^n = a^{m.n} = a^{n.m} = (a^n)^m$ (I have replaced 'x', multiplication symbol, with a 'dot' just for convenience and it is an accepted notation for multiplication).

4.Consider $(2.3)^2$. This is equal to 6^2. We can also write $2^2 x 3^2 = 4x9 = 36$.
Now, $6^2 = 36$. Hence, $(2.3)^2 = 2^2 x 3^2$.
Therefore, $(a.b)^2 = a^2.b^2$
In general, $(a.b)^n = a^n.b^n$

5.Consider,

$$\left(\frac{4^2}{2^2}\right) = \frac{16}{4} = 4$$

$$\left(\frac{4}{2}\right)^2 = 2^2 = 4$$

Therefore, in general,

$$\left(\frac{a^m}{b^m}\right) = \left(\frac{a}{b}\right)^m \left(\frac{a^m}{b^m}\right) = \left(\frac{a}{b}\right)^m$$

6.Square root of 2 is written as $\sqrt{2}$ or $2^{1/2}$. Similarly, cube root of 2 is written as $\sqrt[3]{2}$ or $2^{1/3}$. In general, nth root of a is written as $\sqrt[n]{a} = a^{1/n}$. Also, $\sqrt[q]{a^p} = a^{p/q}$.

7.a^{-m} is written as $\frac{1}{a^m}$.

8.$a^0 = 1$

Examples

1.What is the decimal value of $\left(\frac{1}{8}\right)^2$?

(A) 0.15625 (B) 0.064 (C) 0.015625 (D) 0.0064 (E) 0.015025

Solution:
If you convert 1/8 into decimal fraction, you get 0.125
0.125x0.125 = 0.015625
Therefore, the correct option is (C)
#squaring a number ending with 5. (13x12, 5x5; 156,25; 6 decimal places)

2.if $3^{2x+2}=9^{3x-1}$, what is the value of x?
(A)0 (B)4 (C)1 (D)3 (E)2

Solution:
$3^{2x+2}=\left(3^2\right)^{3x-1}=3^{6x-2}$
Therefore, 2x+2=6x-2
$\Rightarrow x=1$
The correct option is (C)

3.if m is a non-negative integer greater than 0, then $3^m+3^{m+2}=$
(A)$10(3^m)$ (B)30^m (C)3^{2m+2} (D)$3^{m(m+2)}$ (E)$9(3^m)$

Solution:
$3^{m+2}=3^m.3^2$
$\therefore 3^m+3^{m+2}=3^m+3^m.3^2$
Take 3^m outside the bracket (common term).
LHS=$3^m(1+3^2)=3^m(10)$
Therefore, the correct option is (A)

4.What is the value of $(256)^{1/4}+(243)^{1/5}+(343)^{1/3}$?
(A)14 (B)13 (C)12 (D)11 (E)10

Solution:
$(2^8)^{1/4}+(3^5)^{1/5}+(7^3)^{1/3}$
$2^2+3^1+7^1=14$
The correct option is (A)

5.Find the product of $7^{2^3}\times\left(7^2\right)^3\times 7^6$
(A)7^{18} (B)7^{20} (C)7^{19} (D)7^{16} (E)7^{17}

Solution:

$7^{2^3} = 7^8$

$\left(7^2\right)^3 = 7^6$

$\therefore 7^8 \times 7^6 \times 7^6 = 7^{20}$

Therefore, the correct answer is (B)

6.Find the value of $(2^{-1}+3^{-1})^2 \div (2^{-1}-3^{-1})^2$

(A)5 (B)6 (C)0 (D)25 (E)36

Solution:

$\left(\frac{1}{2}+\frac{1}{3}\right)^2 \div \left(\frac{1}{2}-\frac{1}{3}\right)^2$

$\left(\frac{5}{6}\right)^2 \div \left(\frac{1}{6}\right)^2$

$\frac{5^2}{6^2} \div \frac{1^2}{6^2}$

$\frac{5^2}{6^2} \times \frac{6^2}{1^2} = 5^2 = 25$

Therefore, the correct answer is (D)

7.Which of the following is the smallest?

(A)$2^{1/5}$ (B)$3^{1/8}$ (C)$6^{3/40}$ (D)$15^{1/20}$ (E)$5^{1/10}$

Solution:

Number 40 is the L.C.M of the denominators of the exponents.
Take the power of 40.
Now the numbers are 2^8, 3^5,6^3,15^2, and 5^4
They are equal to 256, 243, 216, 225, and 625 respectively.
Out of these, 216 is the smallest number.
Therefore, the correct option is (C).

8.What is the value of $\frac{5^{n+3} - 20 \times 5^{n+1}}{5^{n+1} \times 5}$?

(A)0 (B)1 (C)5 (D)2 (E)3

Solution:

In the numerator, take 5^{n+1} outside the bracket
$5^{n+1}(5^2-20) = 5^{n+1}(5)$
Therefore, the correct option is (B)

9.Which of the following is the largest?
(A)$2^{1/2}$ (B)$3^{1/3}$ (C)$4^{1/4}$ (D)$5^{1/5}$ (E)$6^{1/6}$

Solution:
Look at the options (A) and (C). They are equal. Eliminate A and C.
Compare option B and E
Take 6 as the exponent of the numbers (LCM of 3 and 6)
B is equal to 9 and E is equal to 6. Therefore, B is greater.
Now you have to eliminate one of B and D.
Take 15 as the exponent of the numbers (LCM of 3 and 5)
B is equal to 243 and D is equal to 125.
Therefore, the correct option is (B).
#Instead of comparing all the options simultaneously, it is better to eliminate one by one. Advantage: computation part is easier.

10.What is the value of $(102x98)(102^2-400)$?
(A)100^4 (B)102^4 (C)98^4 (D)100^4+2^4 (E) 100^4-2^4

Solution:
Modify the given expression to suit one of the options.
$(102x98) = (100+2)(100-2) = 100^2-2^2$
$(102^2-400) = (100+2)^2-400 = 100^2+2^2+400-400 = 100^2+2^2$
Now, $(100^2-2^2)(100^2+2^2) = 100^4-2^4$
Therefore, the correct option is (E)

DESCRIPTIVE STATISTICS

The mean is the most used measure of location or average, with the median and the mode being used for applications that are more specific. Each of these statistics has its own characteristics and generally produces a different result for a given set of data. For some sets of data, it is useful to determine all of these statistics but for other data sets, not all three statistics may be valid.

The **arithmetic mean** (usually just shortened to the **mean**) is the name given to the 'simple average' that most people calculate. It is easy to understand and a very effective way of communicating as answer. The **median** is the middle value of an ordered list of data. It is not as well known as the mean but can be more appropriate for certain types of data. The **mode** is the most frequent value or item, typical examples being the most popular model of car or most common shoe size.

Untabulated data will usually be presented to us as a list of numbers. This data can come in any order and can range from a few values to several thousand or more. Suppose we consider just the first ten observations from this data set:

31 9 27 23 10 22 12 24 29 9

To calculate the **mean**, the numbers are added together to find the total and this total is divided by the number of values included. In this case

$$\bar{X} = \frac{31+9+27+23+10+22+12+24+29+9}{10}$$

$$\bar{X} = \frac{196}{10} = 19.6$$

As most statistics require some form of calculation, a notation is developed to describe the necessary steps. Using this shorthand, or notation, the calculation of the mean would be written as follows:

$\bar{X} = \frac{\sum x}{n}$; where x represents individual values, Σ (sigma) is an instruction to sum values, and n is the number of values.

The median is an order statistic (the middle value) and the first step is to rank, or order, the values of observations:

9 9 10 12 22 23 24 27 29 31

The next step is to find the middle value. If the number of observations is odd then we get a single middle value. However, if the number of observations is even then we get two middle values. In the above case, we get 22 and 23 as middle values. The arithmetic mean of these two values is the median. Therefore, the median is 22.5.

The **mode** is the most frequently occurring observation. Given the data:

31 9 27 23 10 22 12 24 29 9

It can easily be seen that 9 occurs twice and is therefore the most frequent value. Therefore, it is the **mode**.

Weighted means

Suppose, in a school there are five divisions in the 5th Grade. We are given the number of students in each division and the average weight of the students in each division. From this data we can find the overall mean of the 5th Grade students. This overall mean is called weighted means.

division	Average($\bar{X}_i$)	No. of students(f_i)	$\bar{X}_i f_i$
A	31	22	682
B	33	20	660
C	34	19	646
D	32	23	736
E	29	25	725
total		109	3449

Weighted means is given by the formula as: $\bar{X} = \frac{\sum \bar{X}_i f_i}{n}$, where n is the total number of students.

$$\bar{X} = \frac{3449}{109} = 31.64$$

The standard deviation

We are sometimes particularly concerned about readings that vary from the expected. In market research, we are interested not only in the typical values but also in whether opinions or behaviours are consistent, or vary considerably. There is a way of measuring this variability or **dispersion**.

The standard deviation is the most widely used measure of dispersion, since it is directly related to the mean. If you choose the mean as the most appropriate measure of central location, then the standard deviation would be the natural choice for a measure of dispersion. The standard deviation measures differences from the mean, a larger value indicating a larger measure of aggregate variation.

We have already seen how to calculate the mean from simple data, and we will need this calculation before we proceed to the calculation of standard deviation. We can again use first 10 observations:
31 9 27 23 10 22 12 24 29 9

The mean of this data is 19.6. The standard deviation is calculated as under:

x	(x- $\bar{X}$)	$(x-\bar{X})^2$
31	11.4	129.96
9	-10.6	112.36
27	7.4	54.76
23	3.4	11.56
10	-9.6	92.16
22	2.4	5.76
12	-7.6	57.76
24	4.4	19.36
29	9.4	88.36
9	-10.6	112.36
196		684.40

$$\bar{X} = \frac{\sum x}{n} = 19.6$$

$$s = \sqrt{\left[\frac{\sum (x-\bar{X})^2}{n}\right]}$$ (Where s is standard deviation and s^2 is called variance)

$$s = \sqrt{\frac{684.40}{10}} = \sqrt{68.44} = 8.27$$

The Range

The range is the most easily understood measure of dispersion as it is the difference between the highest and lowest values. If we were again concerned with the ten observations:

31 9 27 23 10 22 12 24 29 9

the range would be 22 (31-9). It is, however, a rather a crude measure of spread, being dependent on the two extreme observations. It is also highly unstable as new data is added.

Normal distribution

When a variable is continuous, and its value is affected by a large number of chance factors, none of which predominates, then it will frequently have a Normal distribution. The distribution does occur frequently and is probably the most widely used statistical distribution. A Normal distribution is a symmetrical distribution about its mean. Normal distributions come in many shapes and sizes, some will be relatively 'flat', and have a high standard deviation, whilst others will appear 'tall and thin' and have a relatively small standard deviation. These distributions are often summarized by their mean and variance (μ, σ^2 respectively). Normal distributions are characterized particularly by the areas in various sectors of the distribution. It these areas are considered as a proportion of the total area under distribution curve, then they may also be considered as the probabilities of obtaining a value from the distribution in that sector. Theoretically, to find the area under the distribution curve in the sector less than some value x, it is necessary to perform a bit of mathematics every time that you want a probability. Fortunately there is an easier method of finding these areas and hence the associated probabilities.

Looking at the graph of a Normal distribution, we can see that, at least in theory, the values go off to infinity in both directions. If we subtract the mean of the distribution from every value, all we will be doing is shifting the distribution along the axis so that the mean of the new distribution is zero, but the distribution still goes off to infinity in both directions. If we now divide all of the values on the horizontal axis by the standard deviation, we will have a scale, which has a mean of zero and goes off in 'number ofstandard deviations' in either direction. When you do this to any Normal distribution you arrive at something called the **standard Normal distribution.** It is this concept that makes the Normal distribution concept so useful, since, no matter what the variable X represents, and no matter what units it is measured in, we can almost immediately reduce it to this standard Normal distribution.

Therefore, if we define a variable, Z, as the standard Normal variable, we can write it as:

$$Z = \frac{X-\mu}{\sigma}$$

This is known as the transformation of the original variable, and we now find that the areas under this standard Normal distribution are contained in published tables.

1 standard deviation from the mean on both sides of the curve covers 68% of the total area of the curve (Normal distribution curve is bell-shaped and symmetric about the mean; mean equal median).

1.96 standard deviations from the mean on both sides of the curve cover 95% of the total area of the curve.

2 standard deviations from the mean on both sides

2.575 standard deviations from the mean on both sides of the curve cover 99% of the total area of the curve.

3 standard deviations from the mean on both sides of the curve cover 99.7% of the total area of the curve.

Examples

1.If mean is equal to 22/15 times of median and mode is equal to median, then what is the value of x?

X, 3, 4, 5, 5, 5, 7, 11, 21

(A)7 (B)$7^1/_2$ (C)$7^1/_4$ (D)$7^1/_3$ (E)5

Solution:
Mode is equal to median is given
In the given data, mode is 5. Therefore, median is 5.
Now, $\frac{61+x}{9}=\frac{22}{15}\times 5=\frac{22}{3}=\frac{66}{9}$
61+x = 66
x = 5
Therefore, the correct option is (E)

2.What is the average of all possible values of x, if the range of the set{0, 120, -30, 60, x, 60} is 210?
(A)30 (B)35 (C)45 (D)50 (E)25

Solution:
There are three situations.
1)x>120
2)x<-30
3)-30≤x≤120
If x>120, then value of x is x-(-30) = 210; x = 180
If x<-30, then value of x is 120-x = 210; x = -90
The third situation does not arise because the range is 150 in that case.
Therefore, average of 180 and -90 is 45.
Therefore, the correct option is (C).

3.If S is a set of all positive integers such that 6≤x≤14 and S_1 is formed by removing two elements from the set S, the range of S_1 is 7. If the mean of S_1 is $9^4/_7$, what are the two elements of the set S removed?
(A)6, 13 (B) 14, 7 (C)8, 14 (D)9, 14 (E) 6, 12

Solution:
S contains numbers from 6 to 14. Sum of the elements of S = 90
Let sum of the two elements removed as x
Average of the remaining numbers is $\frac{90-x}{7}=\frac{67}{7}$
Therefore, x=23
Since the range of S_1 is 7, one of the elements 6 and 14 must have been removed. If 6 is removed then the other number is 17, that is not possible.
Therefore, one of the elements is 14 and the other is 9.

Hence, the correct option is (D).

4.Find the standard deviation of the ages of four students whose ages are 11, 12, 14, and 15.
(A)4 (B)2 (C)1.5811 (D)2.581 (E)1.6

Solution:
Step 1: Find the mean
The mean is 13; (11+12+14+15)/4

Step 2: Find the squares of the differences between each observed age and the mean age
$(11-13)^2=4$
$(12-13)^2=1$
$(14-13)^2=1$
$(15-13)^2=4$

Step 3: Find the mean of the above squares
(4+1+1+4)/4 = 2.5

Step 4: Find the square root of the above value
$\sqrt{2.5} = 1.5811$

5.The mean score of GMAT score is 529 and standard deviation is 113. If HBS does not select candidates who score less than 98%, then what is the minimum score required for admission into HBS?
(A)760 (B)755 (C)720 (D)740 (E)730

Solution:
The left hand side from the mean of a Normal distribution covers 50% of the total area. Now, we require another 48% on the right hand side of the mean. We know $+\sigma$ covers 34% and $+2\sigma$ covers 48% of the area. Therefore, area below $+2\sigma$ covers 98% of the total area.
Hence, $529+2\sigma$ = 529+2x113=529+226 = 755.
Therefore, the minimum score required is 755. The correct option is (B).

6.The median of the observations 7, 5, 11, 3, 2, x, 16 is given as M. Which of the following is true about M?
(A)$9\leq M\leq 11$ (B)$6\leq M\leq 8$ (C)$7\leq M\leq 9$ (D)$5\leq M\leq 8$ (E)$5\leq M\leq 7$

Solution:
The given set is x, 2, 3, 5, 7, 11, 16

Since the value of x is not known, we have to place x in all positions to find the possible value of the M.
If x is in the first, second, and third position, then the median will be 5.
If x is in the fifth, sixth, and seventh position, then the median will be 7.
If x is in the fourth position, then the value of the median will be between 5 and 7.
In all the above three cases, the value of M is 5 or 7 or between 5 and 7.
Therefore, the correct option is (E)

Functions

The definition of a function consists of two parts.

1.The **rule**: This tells you how values of the function are assigned or calculated.

2.The **domain**: This tells you the set of values to which the rule may be applied.

For example, $f(x)=\sqrt{(x-3)}$, $x\geq3$ defines a function f, where
the rule is $f(x)=\sqrt{(x-3)}$
the domain is $x\geq3$

In describing a domain of a function, the following notation is useful shorthand for defining important sets.

N – the natural numbers 1, 2, 3, . . .
Z – the integers . . . -3, -2, -1, 0, 1 ,2, 3, . . .
Q – the rational numbers (or fractions, including, for example 4/2)
R – the real numbers (both rational and irrational such as $\sqrt{2}$ and π.

This notation can be extended using + and – signs. Thus, R^+ means the positive real numbers; Z^- means the negative integers.
You can also use the symbol $\in$ to mean 'belongs to'. Thus, $x\in R^+$ means 'x belongs to the set of all positive real numbers'.

Note that N is a subset of Z, which is itself a subset of Q, which is a subset of R.

When a function is written down, both the rule and the domain should be given.

Examples

1.If $f(x)=-5x^4+28x^3-33x^2+2x+8$, then find the value of f(1) and f(-1).

Solution:
For f(1), substitute 1 for x in the function
$-5(1^4)+28(1^3)-33(1^2)+2(1)+8$
$-5+28-33+2+8=0$

For f(-1), substitute -1 for x in the function
$-5(-1^4)+28(-1^3)-33(-1^2)+2(-1)+8$
$-5-28-33-2+8=-60$

2.If $f(x)=2x^2-5x$, then find f(x+1).

Solution:
Substitute (x+1) in place of x in the function
$2(x+1)^2-5(x+1)$
$2(x^2+2x+1)-5(x+1)$
$2x^2+4x+2-5x-5$
$2x^2-x-3$
$(2x-3)(x+1)$

3.If $f(x)=\frac{1}{x+1}$, $x\in R$, $x\neq-1$ and $g(x)=x^2+1$, $x\in R$. then, find gof(1/2)
Solution:
First find f(1/2)
We know $f(x)=\frac{1}{x+1}$, $\therefore f(^1/_2)=\frac{1}{^1/_2+1}=\frac{2}{3}$
Now, find g(2/3)
We know $g(x)=x^2+1$
Therefore, $g(2/3)=(^2/_3)^2+1=\frac{4}{9}+1=\frac{13}{9}$
Hence, gof(1/2) = 13/9
#to find fog(1/2), first find g(1/2), and then find f(x), where the value of x is g(1/2).

Rational functions

Your attention in rational functions must always be drawn first to the denominator.

For example, consider the function $\frac{x+1}{x-1}$.
This function is not defined for x=1. The denominator of the fraction equals zero, and division by zero is not defined, so the function can have no value at this point. On the graph of the function this means that there can be no y-value when x=1, and the graph will have a break where x=1. Such a break in the curve is called a **discontinuity.**

Inverse of a function

If a function is such that every value of x is mapped to only one value y and every value of y is mapped to only one value of x, such functions have inverse function.

For example, f(x)=y=2x+3
y=2x+3 implies $x = \frac{y-3}{2}$
Now, interchange x and y; $y = \frac{x-3}{2}$
So, the inverse of f(x) $= \frac{x-3}{2}$

Greatest integer function

The greatest integer function is denoted by y=f(x)=[x].
It denotes the largest integer less than or equal to x.
In general, if m is any integer and x is any real number between m and m+1, that is, m≤x≤m+1, then [x]=m

4.If $x*y=x+y+xy^2-x^2y$, find the value of 3*2.

Solution:
$x+y+xy^2-x^2y=x+y+xy(y-x)$
3*2=3+2+(3x2)(3-2)=5+6x1=11

5.If f(x)=[x] is the greatest integer function. Find the value of f(x) for
(1)x=-0.004 (2)x=0.09

Solution:
(1)The largest integer less than or equal to -0.004 is -1
(2)The largest integer less than or equal to 0.09 is 0

6.If f(x)=3[x] and $g(x)=x^2$, then find fog(x) and gof(x) for x=2.5.

Solution:
fog(2.5)=f(6.25)=3[6.25]=18
gof(2.5)=g(6)=36

7.If f(x)=|x|, g(x, y)=x+y, and h(x, y)=x-y, then find f(g(h(1, 2), -1))

Solution:
f(g(h(1, 2), -1))=f(g(-1, -1))=f(-2)=|-2|=2

Part-II: 101 Selected Problems for Top Score

101-Selected Problems

1.Two cars A and B start from two points P and Q respectively towards each other simultaneously. After traveling a certain distance, at point R, car A develops engine trouble. It continues to travel at two-third of its usual speed to meet car B at a point S where PR=QS. If the engine trouble had occurred after car A had travelled double the distance it would have met car B at a point T where ST=SQ/9. Find the ratio of the speeds of A and B. (Time Speed Distance, Equations, Ratio)

A)4:1 B)2:1 C)3:1 D) 3:2 E)2:3

2.Humpty and Dumpty have a certain number of apples and mangoes between them. The ratio of the number of apples with Humpty to the number of apples with Dumpty is same as that of the ratio of the number of mangoes with Dumpty to the number of mangoes with Humpty. If the number of fruits with Humpty is one more than the number of fruits with Dumpty, then what is the minimum possible number of fruits that they can have between them? (Number properties, Equations, Ratio)

A)15 B)6 C)22 D)9 E)7

3.There are some mangoes and oranges, and there are a certain number of plates. If you place one orange in each plate, one orange is left. If you place two mangoes in each plate, one plate is left without mangoes. If the difference between the number of mangoes and oranges is 3, what is the number of plates? (Number properties, Equations)

A)6 B)9 C)7 D)8 E)cannot be determined

4.What is the number of negative integer pairs (x, y) which satisfy the equation 7x+17y+1000=0? (Number properties, Equations)

(A)7 (B)8 (C)9 (D)10 (E)11

5.In a sports stores, three brands of baseball bats are available. If you buy one bat of each brand, it will cost $1,300. If you buy 2 bats of the first brand, 3 bats of second brand, and 4 bats of third brand, then it will cost $3,110. What is the cost of 5 bats of the first brand, 3 bats of the second brand, and 1 bat of third brand? (Special equations)

(A)$2,320 (B)$4,140 (C)$3,250 (D)$3.920 (E)$5,480

6.The triplets a, b, c; b, c, d; and c, d, e are each in continued proportion. The fourth proportional of a, b, and d is 4. Also, the fourth proportional of a, b, and c is 8. Find the value of a? (Ratio and Proportion)

(A)32 (B)48 (C)24 (D)64 (E)16

7.Set A contains elements which are squares of first 'n' natural numbers such that $n^2 \leq 730$. Set B contains elements which are cubes of first 'm' natural numbers such that $m^3 \leq 730$. What is the sum of all the elements of set A∪B? (Numbers, Sets)

(A)8161 (B)8955 (C)8890 (D)8225 (E)6181

8.Packer had certain number of apples. He always found an apple remained after he packed 2, 3, 4, 5, and 6 apples each time. However, when packed 7 apples each, he found no apple remained. Find the greatest five-digit number of apples he could have. (Number properties)

(A)99,421 (B)99,861 (C)99,961 (D)99,841 (E)99,941

9.In the above problem, if Packer packed 7 or 11 apples he was left with 4 apples each time. If he packed 6 or 9 apples he was left with 2 and 8 apples respectively. If he had less than 1000 apples, how many more apples did he need so that no apple remained when he packed 17 apples each?

(A)6 (B)7 (C)8 (D)9 (E)10

10. $\frac{1}{a}+\frac{1}{b}+\frac{1}{c}+\frac{1}{d}=\frac{1}{210}, a \leq b \leq c \leq d$, what is the largest possible value of d, if a, b, c, and d are positive integers? (Prime factors, Multiple, Fraction, LCM)

(A)1736 (B)1743 (C)1750 (D)1757 (E)1729

11. How many numbers that are not divisible by 15, evenly divide the number 1334025 (Divisors, Prime factors, Sets)?

(A)54 (B)45 (C)36 (D)27 (E)18

12. Which of the following is the largest number that can evenly divide

111^4-7041? (Multiplication, Divisibility)

(A)600000 (B)300000 (C)200000 (D)10000 (E)1000

13. Filly-empty is used to fill as well as to empty a tank whose capacity is 3,600 m^3 in such a way that for the first minute it fills and in the second minute it empties. The pump automatically switches on and off every minute. The pump fills 10 m^3 more in a minute than it empties in a minute. The pump needs 12 more minutes to empty the entire tank alone when it is continuously 'off' mode than to fill it alone when

it is continuously 'on' mode. When the tank is empty at 9 AM, the pump is switched on. At what time will the tank overflow? (Work problems, Progressions)

(A)8.49 PM (B)8.51 PM (C)8.53 PM (D)8.55 PM (E)8.57 PM

14.N_1, N_2, and N_3 are three consecutive positive integers. The sum of the number of divisors of N_1 and N_1^2 is the same as the sum of the number of divisors of N_2 and N_2^2, and N_3 and N_3^2. Which of the following is the greatest number that can evenly divide the product of N_1,N_2, and N_3, if they are the first three consecutive positive integers which satisfy the above condition?

(A)195 (B)102 (C)221 (D)119 (E)187

15.N, the set of natural numbers, is divided into subsets A_1=(1), A_2=(2, 3), A_3=(4, 5, 6), A_4=(7, 8, 9, 10) and so on. Find the mean of the elements of A_{49}. (Numbers, Descriptive statistics)

(A)1105 (B)1250.5 (C)1201 (D)1152.5 (E)1301

16.A four digit number is in the form of aabb, such that 'a' and 'b' are distinct and a>0. The number is divisible by 121. What could be the value of 'a', if the number has the greatest number of factors?

(A)4 (B)5 (C)6 (D)7 (E)8

17.How many pairs of positive integers m, n satisfy $\frac{1}{m}+\frac{4}{n}=\frac{1}{12}$, where n is less than 56? (Equations, Number properties)

(A)4 (B)7 (C)5 (D)3 (E)6

18.The sum of 5 consecutive two digit even numbers, when divided by 10, becomes a perfect square. Which of the following cannot be one of those 5 numbers? (Number properties, Equations)

(A)76 (B)70 (C)54 (D)64 (E)36

19.A travels from X to Y at the speed of 60 mph. While A starts his journey from X at 3 PM, B starts her journey at 3.50 PM. They cross each other at 4 PM. The total journey-time of A and B together is 132 minutes and A reached Y before B reached X. Find the distance between X and Y. (Distance, Speed, Time)

A)110 miles B)132 miles C)72 miles D)82 miles E)92miles

20.Find the value of $\frac{1}{2!}+\frac{2}{3!}+\frac{3}{4!}+\frac{4}{5!}+\frac{5}{6!}+\frac{6}{7!}$(Series)

A)$\frac{7!-1}{7!}$B)$\frac{6!+1}{7!}$C)$\frac{6!-1}{7!}$D)$\frac{21}{7!}$E)$\frac{6}{7}$

21.A watermelon weighed 600 pounds. Ninety nine percent of its weight was due to the water content in the watermelon. After it was kept in a drying room for a while, it turned out that it was only 98% water by weight. How much does it weigh now? (Percentage, Proportions)

A)588 pounds B)594 pounds C)294 pounds D)300 pounds E)590 pounds

22.Nine teams play in a tournament. Each team plays every other team exactly once. A team receives 3 points for a win, 2 points for a draw, and 1 point for a loss. Every team ends up with a different total score and the difference between any two successive scores is same when they are ranked. The team with the highest total scored 20. What is the arithmetic mean score of the possible scores under loss that the team scored the lowest total? (Combinations, Progressions, and Average)

A)6 B)5 C)4 D)3 E)2

23.If the equation $(p+4)x^2-2px+2p-6=0$ is satisfied for all real values of x, the range of values of p is: (Equations)

A)[-6,4] B) (-6,4) C) [-6,4) D) (-6,4] E) $-6<p\leq 4$

24.Find the ratio of total surface area of a cube to the largest tetrahedron, whose edges are equal, that could be kept inside the cube. (Solid geometry)

A)$2:\sqrt{3}$ B)$3:2\sqrt{3}$ C)$2\sqrt{3}:3$ D)$2\sqrt{3}:1$ E)$\sqrt{3}:1$

25.What is the volume of the largest cube that can be kept inside a right circular cone whose radius and height are 1 unit and 3 units respectively? (Solid geometry, Similar triangle properties)

A)$\frac{36}{22+12\sqrt{2}}$ B)$\frac{108}{58+45\sqrt{2}}$ C)$\frac{54}{34+30\sqrt{2}}$ D)$\frac{36}{58+45\sqrt{2}}$ E)$\frac{54}{22+12\sqrt{2}}$

26.A right triangle has sides of length l, m, 27. Either l or m is the hypotenuse of the triangle. What is the arithmetic mean of all possible values of l and m? (Plane geometry and Average)

A)172.5 B)173.5 C)174.5 D)175.5 E)176.5

27.Suppose that you write 11 letters and then you address 11 envelopes to go with them. Closing your eyes, you randomly stuff one letter into each envelope. What is the probability that precisely two letters are in the wrong envelopes and all others in the correct envelopes? (Probability)

A)$\frac{1}{9!}$B)$\frac{1}{2\times11!}$C)$\frac{1}{11!}$D)$\frac{1}{2\times10!}$E)$\frac{1}{2\times9!}$

28.Eight slips of paper with the numbers 1, 2, 3, 4, 5, 6, 7, and 8 written on them are placed into a bin. The eight slips are drawn one by one from the bin. What is the probability that the first four to come out are 1, 3, 5, 7 in some order? (Probability)

A)1/70 B)1/14 C)1/35 D)1/140 E)1/2

29.A standard deck of 52 playing cards are divided into 4 sub-decks and kept on a table face down. What is the probability that one of those four top cards is a face card? (Probability)

A).396 B).553 C).496 D).453 E).353

30.Draw a planer grid that is 31 squares wide and 17 squares high. Then you remove squares so that the new grid is a 17 by 17-square grid. If a and b are the number of rectangles that can be drawn using the lines of the 31 by17 grid and 17 by 17 grid respectively then find (a-b).

A)52379 B)52479 C)51379 D)51479 E)50479

31.A pyramid with a slant height of 4 units in all its slant edges is cut at one corner of a cube with a side length of 8 units. Find the volume of the pyramid. (Solid geometry)

A)10.17 B)10.97 C)10.37 D)10.67 E)10.27

32.Let B1 and B2 be two boxes such that B1 contains 3 white and 2 red balls and B2 contains only 1 white ball. A fair coin is tossed. If head appears then 1 ball is drawn at random from B1 and put into B2. However, if tail appears then 2 balls are drawn at random from B1 and put into B2. Now 1 ball is drawn at random from B2. Find the probability that the drawn ball from B2 being white. (Probability)

A)13/30 B)23/30 C)19/30 D)11/30 E)1/2

33.A and B throw 2 dice at a time, one by one. If sum is 8 then thrower wins. A starts the game and if probability of his winning is x/(2x-5) find the value of x. (Probability, Progressions)

A)36 B)67 C)-36 D)-67 E)-25/26

34.A two digit number is 6 more than another two digit number. Then the first number is written in base 5 and the second number is written in base 4. It happened that the second number is the reverse of the first number. Find the product of these two numbers in base 10. (Number properties)

A)156 B)180 C)165 D)187 E)209

35.B&W Cargo charges $16000 as the freight charges when the distance to be covered is 1000 miles and $10000 for a distance of 550 miles. Find the freight charge for 280 miles. (Ratio and proportions)

A)$5200 B)$5100 C)$5900 D)$6100 E)$6400

36.An empty tank is half filled with milk. Then it is filled with water up to half of the remaining. Then it is filled with milk up to half of the remaining. The process continues until the tank is filled. Find the ratio of milk to water. (Ratio, Progressions)

A)1:1 B)3:1 C)2:1 D)4:1 E)3:2

37.A can do a piece of work in 32 days. A and B together can do the same piece of work in 8 days. They undertook a work, which they can complete in 40 days, if they worked together, for $16000. For the first 10 days, B was absent and then they completed the remaining part working together. In how many days did they complete the whole work? (Work and Time)

A)50 days B)47.5 days C)52.5 days D)52 days E)42.5 days

38.In the above question, what is the ratio of the remuneration of A to B?

A)10:33 B)19:45 C)17:45 D)1:3 E)2:5

39.A man bought two tables for $1440. He sold one of them at a gain of 15% and the other at a loss of 8%. It was then found that each table fetched him the same amount. Find the cost price of second table. (Profit and Loss, Ratio)

A)$800 B)$740 C)$940 D)$640 E)$880

40.If x travels at 30 miles per hour, then x reaches the destination late by 10 minutes. If x travels at 42 miles per hour, then x reaches the destination 10 minutes early. Find the distance. (Time, Speed, and Distance)

A)30 miles B)42 miles C)38 miles D)35 miles E)40 miles

41.X, Y, and Z are three typists who working simultaneously can type 216 pages in four hours. In one hour X can type as many pages more than Y as Y can type more than Z. During a period of five hours, X can type as many pages as Z can during seven hours. How many pages does Y type per hour? (Work, Ratio)

A)15 B)16 C)17 D)18 E)19

42.A can do a piece of work in 18 days, B in 9 days and C in 6 days. A and B start working together and after 2 days C joins them. What is the total number of days taken to finish the work? (Work, Time)

A)6 B)5 C)4 D)3.5 E)3

43.A is twice as good a workman as B and finished a piece of work in t hours less than B. In how many hours they together could finish that piece of work. (Work, Time)

A)3t/2 B)2t/3 C)2t D)3t E)4t/3

44.From a barrel containing 500 liters of milk, 3 mugs of milk are poured into a barrel containing 1000 liters of water. After mixing the contents well, 3 mugs of the mixture are poured into the barrel of milk. What is the ratio of the percentage of water in the barrel of milk to the percentage of milk in the barrel of water? (Mixture, Ratio, Percentage)

A)1:1 B)2:1 C)1:2 D)2:3 E)3:2

45.An alloy contains copper and zinc in the ratio 5:3 and another allow contains copper and tin in the ratio 8:5. If equal quantities of both the alloys are melted together to form another alloy then what is the percentage of tin in the resultant alloy. (Mixture, Percentage)

A)500/21 B)200/13 C)150/13 D)175/13 E)250/13

46.If $1066 are divided among A,B,C, and D such that A:B=3:4; B:C=5:6; and C:D=7:5. Who will get the maximum? (Ratio and Proportions)

A)A B)B C)C D)D E)E

47.The total emoluments of A and B are equal. However, A gets 65% of his basic salary as allowances and B gets 80% of his basic salary as allowances. What is the ratio of the basic salaries of A and B? (Ratio, Percentage)

A)11:12 B)16:13 C)12:11 D)13:16 E)13:12

48.A family wanted to reduce the expenditure on milk by 19%. However, the price of milk increased by 8%. In order to achieve the desired result, by what percentage the family would reduce their consumption? (Percentage)

A)20% B)27% C)26.2% D)23.75% E)25%

49.A and B run a 1400 meter race. In the first heat, A gives B a start of 35m and beats B by 40 seconds. In the second heat, A gives B a start of 60 seconds and is beaten by 40 meter. In what time could A run a 2 km race? (Race problems)

A)500 sec B)400 sec C)450 sec D)480 sec E)350 sec

50.In what scale notation is 25 doubled by reversing the digits.

A)5 B)6 C)7 D)8 E)9

51.A waterman rows to a place 48 miles distance and back in 14 hours: he finds that he can row 4 miles with the stream in the same time as 3 miles against the stream. Find the rate of the stream. (Time, Speed, Distance)

A)7 mph B)5 mph C)3 mph D)2 mph E)1 mph

52.A family consumes 20 loaves of bread in a week. If wages were raised 5 percent and the price of bread were raised 2.5%, the family would gain 30 cents a week. However, if wages were lowered 7.5% and bread fell 10%, then the family would lose 10 cents a week. Find the price of a loaf. (Percentage, Equations)

A)32 cents B)30 cents C)28 cents D)26 cents E)24 cents

53.A train, an hour after starting, meets with an accident which detains it an hour, after which it proceeds at three-fifths of its former speed and arrives 3 hours late. However, had the accident happened 100 km farther on the line, it would have arrived 1.5 hours sooner. Find the length of the journey. (Time, Speed, Distance)

A)1600/9 km B)1500/9 km C)1400/9 km D)1300/9 km E)1200/9 km

54.In what time A,B, and C together do a work if A alone could do it in six hours more, B alone in one hour more, and C alone in twice the time? (Work and Time, Quadratic equations)

A)70 min B)60 min C)50 min D)40 min E)30 min

55.A vessel contains 15 liters of wine and another vessel contains 10 liters of water: 6 liters are taken out of each vessel and transferred to the other; this operation is repeated for 4 times. Find the percentage of wine in the first vessel after the fourth operation. (Mixture, Ratio, and Percentage)

A)20% B)30% C)40% D)50% E)60%

56.In a certain community consisting of p persons, a percent can read and write; of the males alone b percent, and of the females alone c percent can read and write. Find the number of females in the community if p=900, (a-c)=10, and (b-c)=15. (Percentage, Equations)

A)300 B)400 C)500 D)600 E)200

57.The interior angles of a pentagon are in a ratio of an arithmetic progression whose first term is 4 and common difference is 1. Find the difference between the largest and the smallest angle. (Geometry, Ratio, Progressions)

A)108 B)71 C)54 D)72 E)60

58.The complement of an angle exceeds the angle itself by 40°. Find the angle.

A)70° B)65° C)60° D)55° E)50°

59.The supplement of an angle is one-fifth of the angle itself. Find the angle.

A)130° B)140° C)150° D)160° E)120°

60.In a triangle ABC, $AB^2+AC^2=200$ cm^2. Median, AD=8 cm. find BC. (Properties of triangles)

A)16 B)15 C)$10\sqrt{2}$ D)$9\sqrt{2}$ E)12

61.The exterior angle of a regular polygon is one-third of its interior angle. How many sides has the polygon?

A)9 B)8 C)7 D)6 E)5

62.Two regular polygons are such that the ratio between their number of sides is 1:2 and the ratio of measures of their interior angles is 3:4. Find the interior angle of the first polygon.

A)140° B)135° C)120° D)108° E)90°

63.One of the angles of a triangle is 144°. Find the angle between the bisectors of the acute angles of the triangle.

A)162° B)144° C)126° D)118° E)108°

64.D and E are the middle points of the sides AB and AC of a triangle ABC. Find the ratio of the area of triangle ADE to area of triangle ABC.

A)1:2 B)1:3 C)1:4 D)2:3 E)3:5

65.In a triangle ABC, the base angles B and C are 45° each. D is mid-point on BC. AD=50 cm. Find the length of BC.

A)$50\sqrt{2}$ cm B)$50/\sqrt{2}$ cm C)50 cm D)$100/\sqrt{3}$ cm E)100 cm

66.ABC is right angled triangle at A. Angle B is twice angle C. Which of the following is true?

A)$AC^2=2AB^2$ B)$2AC^2=AB^2$ C)$3AC^2=AB^2$ D)$AC^2=3AB^2$ E)$AC^2=4AB^2$

67.The present value of $672 due in a certain time is $126. If compounded interest at $4\frac{1}{6}\%$ be allowed, find the time.

A)17 years B)25 years C)35 years D)41 years E)53 years

68.At simple interest the interest on a certain sum of money is $90, and the discount on the same sum for the same time and at the same rate is $80. Find the sum.

A)$640 B)$660 C)$680 D)$700 E)$720

69.A freehold estate is bought for $275,000; at what rent per annum should be let so that the owner may receive 4 per cent per annum on the purchase money allowing compound interest?

A)$11,000 B)$10,500 C)$10,000 D)$9,500 E)$9,000

70.Three travelers arrive at a town where there are four hotels; in how many ways can they take up their quarters, each at a different hotel?

A)6 B)12 C)16 D)18 E)24

71.In how many ways can 15 recruits be divided into three equal groups?

A)126126 B)252252 C)378378 D)504504 E)756756

72.The duration of a railway journey varies directly as the distance and inversely as the velocity; the velocity varies directly as square root of the quantity of coal used per mile, and inversely as the number of carriages in the train. In a journey of 50 miles in half an hour with 18 carriages 100 kg of coal is consumed; how much coal will be consumed in a journey of 42 miles in 28 minutes with 16 carriages? (Variation)

A)64.76 B)80.76 C)53.76 D)52.76 E)51.76

73.A man starts from A to B, another starts from B to A at the same time. After they meet, they complete their journeys in 10/7 and 14/5 hours respectively. Find the speed of the second man if the speed of first is 70 miles per hour. (Time, Speed, and Distance)

A)65 mph B)60 mph C)55 mph D)50 mph E)45 mph

74.I shall be 40 minutes late to reach my office if I walk from my house at 3 miles per hour. I shall be 30 minutes early if I walk at 4 miles per hour. Find the distance between my house and the office.

A)10 miles B)11 miles C)12 miles D)13 miles E)14 miles

75.A and B walk from P to Q a distance of 21 miles at 3 and 4 miles per hour respectively. B reaches Q and immediately returns and meets A at R. Find the distance from P to R.

A)14 miles B)15 miles C)16 miles D)17 miles E)18 miles

76.A train after travelling 50 miles from A meets with an accident and proceeds at four-fifths of the former speed and reaches B, 45 minutes late. Had the accident happened 20 miles further on, it would have arrived 12 minutes sooner. Find the distance between A and B.

A)100 miles B)125 miles C)140 miles D)150 miles E)175

77.A train leaving L at 3.10 p.m. reaches W at 5.00 p.m. One leaving W at 3.30 p.m. arrives in L at 5.50 p.m. When do they pass each other?

A)16:18:24 B)16:19:24 C)16:20:24 D)16:21:24 E)16:22:24

78.If x<1, find the sum of the series $1+2x+3x^2+4x^3+....\infty$

A)$\frac{1}{1-x}$ B)$\frac{1}{(1-x)^2}$ C)$\frac{x}{1-x}$ D)$\frac{x}{(1-x)^2}$ E)$\frac{1-x}{x}$

79.Find the sum to infinity of the series:

$$1+\frac{4}{5}+\frac{7}{5^2}+\frac{10}{5^3}+\cdots$$

A)16/35 B)17/35 C)35/17 D)35/16 E)35/18

80.The arithmetic and geometric mean of two numbers are 6.5 and 6 respectively. Find the harmonic mean of the two numbers.

A)5.938 B)5.838 C)5.738 D)5.638 E)5.538

81.Find the sum of the series, 1x2+2x3+3x4+...., to 20 terms.

A)1540 B)3080 C)2540 D)1580 E)3040

82.Find the minimum value of x^2-4x+7 for real values of x.

A)2 B)-2 C)3 D)-3 E)1

83.If x is positive, find the greatest value of (5-x)(x+3).

A)20 B)19 C)18 D)17 E)16

84.Find the solution set of x if $|2x-5|<4x+9$

A)$x<2/3$ B)$x<-2/3$ C)$x>2/3$ D)$x>-2/3$ E)$x>-7$

85.Find the area of the quadrilateral formed by the solution set of the inequalities $2x+3y\leq12$, $x\geq0$, $y\geq0$ and $x\leq3$.

A)9 B)8 C)7 D)6 E)5

86.The perimeter of a rectangle is 100 cm. Find the length of its sides when its area is maximum.

A)45, 5 B)40, 10 C)35, 15 D)30, 20 E)25, 25

87.The total cost of producing 'x' machines is $f(x)=2000+100x-0.1x^2$. Find the average cost for producing 100 machines.

A)130 B)120 C)110 D)100 E)90

88.If the price of sugar is increased by 15% find how much percent a householder must reduce her consumption of sugar so as not to increase the expenditure.

A)15% B)10% C)$11\frac{1}{23}\%$ D)$12\frac{1}{23}\%$ E)$13\frac{1}{23}\%$

89.If the price of sugar is decreased by 15 percent find how much percent a householder can increase her consumption on the same budget.

A)$15\frac{11}{17}\%$ B)$16\frac{11}{17}\%$ C)$17\frac{11}{17}\%$ D)$18\frac{11}{17}\%$ E)$19\frac{11}{17}\%$

90.The value of a machine depreciates at the rate of 10% per annum. If its present value is $81,000, what was the value of the machine 2 years ago?

A)$98,010 B)$90,000 C)$99,000 D)$100,000 E)$99,010

91.The average weight of a class of 24 students is 35 kilo gram. If the weight of the teacher is included, the average rises by 400 grams. Find the weight of the teacher.

A)45 kg B)44 kg C)43 kg D)42 kg E)41 kg

92.A batsman makes a score of 87 runs in the 17th inning and thus increased his average by 3. Find the average after 17th inning.

A)36 B)37 C)38 D)39 E)40

93.The average age of a committee of 8 members is 40 years. A member aged 55 years retired and a new member aged 39 joined in his place. Find the average age of the present committee.

A)37 B)38 C)39 D)40 E)41

94.A dishonest dealer professes to sell his goods at cost price but uses a weight of 960 grams for 1 kg. Find his gain percent.

A)4% B)$4\frac{1}{3}\%$ C)$4\frac{1}{4}\%$ D)$4\frac{1}{5}\%$ E)$4\frac{1}{6}\%$

95.750 men have provisions for 20 days. If at the end of 4 days, 450 men join the existing force, how long will the remaining provision last?

A)10 days B)11 days C)12 days D)13 days E)14 days

96.If x men working x hours can produce x units in x days, then how many units can y men working y hours in y days produce?

A)x^3/y^3 B)x^3/y^2 C)y^3/x^3 D)y^3/x^2 E)y

97.A sum was put at simple interest at a certain rate for 2 years. Had it been put at 3% higher rate, it would have fetched $300 more. Find the sum.

A)$2,000 B)$2,500 C)$3,500 D)$4,500 E)$5,000

98.What annual installment will discharge a debt of $2,210 due in 4 years at 7% simple interest?

A)$450 B)$475 C)$500 D)$525 E)$550

99.The difference between compound interest and simple interest on a certain sum at 10% p.a. for 2 years is $52. Find the sum.

A)$5,200 B)$5,100 C)$5,000 D)$5,300 E)$5,400

100.In what ratio must a person mix three kinds of wheat costing him $1.20, $1.44, and $1.74 per kg so that the mixture may be worth $1.41 per kg?

A)7:7:12 B)7:12:7 C)12:7:7 D)12:9:7 E)12:9:9

101.A and B solved a quadratic equation. In solving it, A made a mistake in the constant term and got the roots as 6 and 2, while B made a mistake in the coefficient of x only and got the roots as -7 and -1. Find the correct roots.

A)(7, 1) B)(2, 7) C)(6, -1) D)(6, -7) E)(2, -1)

ANSWERS WITH EXPLANATION

1.Two cars A and B start from two points P and Q respectively towards each other simultaneously. After traveling a certain distance, at point R, car A develops engine trouble. It continues to travel at two-third of its usual speed to meet car B at a point S where PR=QS. If the engine trouble had occurred after car A had travelled double the distance it would have met car B at a point T where ST=SQ/9. Find the ratio of the speeds of A and B. (Time Speed Distance, Equations, Ratio)

A)4:1 B)2:1 C)3:1 D) 3:2 E)2:3

Answer: Option (C)

Let PR=RS=QS=x (RS can be 'y' or 'x', it does not make any difference)

Let the usual speeds of A and B are 's' and 's1' respectively

Case 1: $\frac{x}{s}+\frac{3x}{2s}=\frac{x}{s1}$---equation 1

Case 2: $\frac{2x}{s}+\frac{3x}{18s}=\frac{8x}{9s1}$ ---equation 2

1-2 gives $\frac{s}{s1}=\frac{3}{1}$

2.Humpty and Dumpty have a certain number of apples and mangoes between them. The ratio of the number of apples with Humpty to the number of apples with Dumpty is same as that of the ratio of the number of mangoes with Dumpty to the number of mangoes with Humpty. If the number of fruits with Humpty is one more than the number of fruits with Dumpty, then what is the minimum possible number of fruits that they can have between them? (Number properties, Equations, Ratio)

A)15 B)6 C)22 D)9 E)7

Answer: Option (D)

From the second sentence, we know that $H_a.H_m=D_a.D_m$ - equation(1)

From the third sentence, we know that $H_a+H_m=D_a+D_m+1$- equation(2)

From the second equation, the total number of fruits with Humpty and Dumpty is odd.

Our first choice is option D. The number of fruits with Humpty and Dumpty is 5 and 4 respectively.

According to the condition of first equation, the only possibility is 4x1=2x2.

$$\frac{4}{2}=\frac{2}{1}$$

Number of apples and mangoes with Humpty is 4 and 1 respectively.

Number of apples and mangoes with Dumpty is 2 and 2 respectively.

3.There are some mangoes and oranges, and there are a certain number of plates. If you place one orange in each plate, one orange is left. If you place two mangoes in each plate, one plate is left without mangoes. If the difference between the number of mangoes and oranges is 3, what is the number of plates? (Number properties, Equations)

A)6 B)9 C)7 D)8 E)cannot be determined

Answer: Option (A)

From sentence three, you know that number of mangoes is even.

From sentence four, you know that number of oranges is odd.

From sentence two, you know that number of plates is even.

Our first choice is option (A)

Plates:6, mangoes:10, and oranges:7; the difference is 3.

Plates:8, mangoes:14, and oranges:9; the difference is 5.

4.What is the number of negative integer pairs (x, y) which satisfy the equation 7x+17y+1000=0? (Number properties, Equations)

(A)7 (B)8 (C)9 (D)10 (E)11

Answer: option (C)

$$x=\frac{-1000-17y}{7}$$

$$x=\frac{-1001-14y-3y+1}{7}$$ (-1001-14y is divisible by 7; we have to check only-3y+1)

y= (-2, -9, -16, -23, -30, -37, -44, -51, -58)

5.In a sports stores, three brands of baseball bats are available. If you buy one bat of each brand, it will cost $1,300. If you buy 2 bats of the first brand, 3 bats of second brand, and 4 bats of third brand, then it will cost $3,110. What is the cost of 5 bats of the first brand, 3 bats of the second brand, and 1 bat of third brand? (Special equations)

(A)$2,320 (B)$4,140 (C)$3,250 (D)$3.920 (E)$5,480

Answer: option (E)

x+y+z = 1300 -equation (1)

2x+3y+4z = 3110 -equation (2)

We have to find 5x+3y+z

If you multiply the first equation you get 5x+5y+5z = 6500

As it has (2y+4z) more, you have to eliminate it.

Multiply the first equation with 2 and subtract it from equation 2. You get y+2z = 510. Therefore, 2y+4z = 1020

Hence, the required answer is 6500-1020 = $5,480

6.The triplets a, b, c; b, c, d; and c, d, e are each in continued proportion. The fourth proportional of a, b, and d is 4. Also, the fourth proportional of a, b, and c is 8. Find the value of a? (Ratio and Proportion)

(A)32 (B)48 (C)24 (D)64 (E)16

Answer: option (D)

From sentence 2, we know, $\frac{a}{b} = \frac{d}{4}$---(1)

From sentence 3, we know, $\frac{a}{b} = \frac{c}{8}$---(2)

Combining the above two equations, we get c=2d ---(3)

From the first sentence, we know, $\frac{a}{b} = \frac{b}{c} = \frac{c}{d} = \frac{d}{e} = \frac{2}{1}$(since c=2d) ---(4)

From equation (2), we get c = 16 (since a:b::2:1), which implies b = 32 and a = 64

7.Set A contains elements which are squares of first 'n' natural numbers such that $n^2 \leq 730$. Set B contains elements which are cubes of first 'm' natural numbers such that $m^3 \leq 730$. What is the sum of all the elements of set A∪B? (Numbers, Sets)

(A)8161 (B)8955 (C)8890 (D)8225 (E)6181

Answer: option (A)

A = $\{1^2, 2^2, 3^2, 4^2, \quad ,27^2\}$

B = $\{1^3, 2^3, 3^3, 4^3, \quad ,9^3\}$

There are some elements, which are common in set A and set B (1, 64, and 729).

Using the formula for sum of squares of first 'n' natural numbers and sum of cubes of first 'm' natural numbers, we get

$$\left(\frac{9\times10}{2}\right)^2+\left(\frac{27\times28\times55}{6}\right)-(1+64+729)=2025+6930-794=8161$$

8.Packer had certain number of apples. He always found an apple remained after he packed 2, 3, 4, 5, and 6 apples each time. However, when packed 7 apples each, he found no apple remained. Find the greatest five-digit number of apples he could have. (Number properties)

(A)99,421 (B)99,861 (C)99,961 (D)99,841 (E)99,941

Answer: option (D)

From the second sentence, the generating term is 60k+1 (LCM of 2, 3, 4, 5, and 6 is 60)

From the third sentence, the generating term is $7k_1$

Therefore, 60k+1 = $7k_1$

Now we have to find the cursive generating term, which satisfies both conditions.

We know, $k_1=\frac{60k+1}{7}$

It implies 'k' takes the values 5, 12, 19, and so on

Therefore, 'k' is in the form of (7n-2)

Substituting (7n-2) in 60k+1, we get, 420n-119

This is the number we require. First, find the greatest five-digit number, which is divisible by 420 and then subtract 119 from that number.

99,960 is the greatest five-digit number divisible by 420. Hence, the required number is 99,960-119 = 99841

9.In the above problem, if Packer packed 7 or 11 apples he was left with 4 apples each time. If he packed 6 or 9 apples he was left with 2 and 8 apples respectively. If he had less than 1000 apples, how many more apples did he need so that no apple remained when he packed 17 apples each?

(A)6 (B)7 (C)8 (D)9 (E)10

Answer: option (D)

From the first sentence, we have 77k+4<1000

From the second sentence, we have $6k_1+2=9k_2+8<1000$

Using the cursive generating term in $6k_1+2=9k_2+8$, we know k_1 takes a generating term (3n+1)

Therefore, $6k_1+2=6(3n+1)+2=18n+8$

Now, 77k+4 = 18n+8

Using the cursive generating term again, we know k takes a generating term (18m-10)

Substituting in 77k+4, we get 1386m-766<1000

When m=1, the first term is 620 which is less than 1000

If you add 9 to 620 then it is divisible by 17. Therefore, the required answer is 9.

10.$\frac{1}{a}+\frac{1}{b}+\frac{1}{c}+\frac{1}{d}=\frac{1}{210}, a \leq b \leq c \leq d,$ what is the largest possible value of d, if a, b, c, and d are positive integers? (Prime factors, Multiple, Fraction, LCM)

(A)1736 (B)1743 (C)1750 (D)1757 (E)1729

Answer: option (E)

210=2x3x5x7

Now, $\frac{1}{2}+\frac{1}{3}+\frac{1}{5}+\frac{1}{7}=\frac{247}{210}$

Divide both sides by 247, then the last term, $\frac{1}{d}=\frac{1}{1729}$

Therefore, d=1729

11.How many numbers that are not divisible by 15, evenly divide the number 1334025 (Divisors, Prime factors, Sets)

(A)54 (B)45 (C)36 (D)27 (E)18

Answer: option (B)

$1334025=3^2 x 5^2 x 7^2 x 11^2$

In general, when $N=a^p x b^q x c^r$, where a, b, and c are the prime factors of N and p, q, and r are positive integer exponents of a, b, and c respectively, the number of divisors of N = (p+1)(q+1)(r+1).

Now, the number is not divisible by 15. Therefore, we should exclude one of the prime factors 3 and 5 from our calculation. We have two cases.

a)$3^2 x 7^2 x 11^2$

b)$5^2 x 7^2 x 11^2$

Number of divisors in the first case is 27 and in the second case is also 27.

In this, the divisors of $7^2 x 11^2$ are included twice. Therefore, we have to subtract 9 from 54 to get the answer. The correct option is (B)

12.Which of the following is the largest number that can evenly divide

111^4-7041? (Multiplication, Divisibility)

(A)600000 (B)300000 (C)200000 (D)10000 (E)1000

Answer: option (A)

If you know multiplication rule of 111 you can do it quickly. Otherwise, it might take more time.

$111^2 = 12321$

1

1+1=2

1+1+1=3

1+1=2

1

$111^3 = 1367631$

1

1+2=3

1+2+3=6

2+3+2=7

3+2+1=6

2+1=3

1

Hope you got it.

111^4=151807041

When you subtract 7041, you get 151800000. 6 can evenly divide 1518. Therefore, the correct option is (A)

13.Filly-empty is used to fill as well as to empty a tank whose capacity is 3,600 m^3 in such a way that for the first minute it fills and in the second minute it empties. The pump automatically switches on and off every minute. The pump fills 10 m^3 more in a minute than it empties in a minute. The pump needs 12 more minutes to empty the entire tank alone when it is continuously 'off' mode than to fill it alone when it is continuously 'on' mode. When the tank is empty at 9 AM, the pump is switched on. At what time will the tank overflow? (Work problems, Progressions)

(A)8.49 PM (B)8.51 PM (C)8.53 PM (D)8.55 PM (E)8.57 PM

Answer: option (B)

Let x be the emptying capacity of the pump and t be the time taken by the pump to fill the tank alone when it is completely 'on' mode.

$$\frac{3600}{x+10} = t; \frac{3600}{x} = t + 12$$

Solving, we get x=50 m^3/min and t=60 minutes

At the end of 1 min, the tank is 60 m^3 full

At the end of 3 min, the tank is 70 m^3 full

Setting progression, at the end of 709 minutes, the tank is 3,600 m^3 full

At the end of 710 minutes, the tank is 3,550 m^3 full

Between 710 and 711 minutes, the tank will overflow

It takes 11 hours and 51 minutes.

14.N_1, N_2, and N_3 are three consecutive positive integers. The sum of the number of divisors of N_1 and N_1^2 is the same as the sum of the number of divisors of N_2 and N_2^2, and N_3 and N_3^2. Which of the following is the greatest number that can evenly divide the product of N_1,N_2, and N_3, if they are the first three consecutive positive integers which satisfy the above condition?

(A)195 (B)102 (C)221 (D)119 (E)187

Answer: option (E)

We need three consecutive integers, which have the same number of divisors. The numbers could be either three consecutive prime numbers or product of two prime numbers. The first such set is 33, 34, and 35.

33=3x11; the number of divisors=4

34=2x17; the number of divisors=4

35=5x7; the number of divisors=4

Now, 33x34x35 is divisible by 187 (11x17)

15.N, the set of natural numbers, is divided into subsets A_1=(1), A_2=(2, 3), A_3=(4, 5, 6), A_4=(7, 8, 9, 10) and so on. Find the mean of the elements of A_{49}. (Numbers, Descriptive statistics)

(A)1105 (B)1250.5 (C)1201 (D)1152.5 (E)1301

Answer: option (C)

The set of means of A_1, A_2, A_3, and so on=(1, 2.5, 5, 8.5, 13, and so on)

The generating term of the sequence of the means is $\frac{n^2+1}{2}$

Therefore, the mean of A_{49}= $(49^2+1)/2 = 1201$

16.A four digit number is in the form of aabb, such that 'a' and 'b' are distinct and a>0. The number is divisible by 121. What could be the value of 'a', if the number has the greatest number of factors?

(A)4 (B)5 (C)6 (D)7 (E)8

Answer: option (D)

The four digit number can be written as 1000a+100a+10b+b=1100a+11b

The four digit number 11(100a+b) is divisible by 11. However, it is given that the number is divisible by 121. That implies (100a+b) is divisible by 11 or (a+b) is divisible by 11. The pairs which satisfy the condition are: (2, 9), (9, 2), (3, 8), (8, 3), (4, 7), (7, 4), (5, 6), and (6, 5).

The possible values of 100a+b: 209, 902, 308, 803, 407, 704, 506, and 605

Out of these number, 704 = 64x11

$64 = 2^6$

The number is $11^2 x 2^6$, which gives the greatest number of factors out of the given set. Hence, a=7 and b=4.

17.How many pairs of positive integers m, n satisfy $\frac{1}{m}+\frac{4}{n}=\frac{1}{12}$, where n is less than 56? (Equations, Number properties)

(A)4 (B)7 (C)5 (D)3 (E)6

Answer: option (C)

It is given that n<56, and we know that n is not less than or equal to 48.

Therefore, n takes values from 49 to 55. There are seven numbers.

$$\frac{1}{m}=\frac{1}{12}-\frac{4}{n}$$

$$\frac{1}{m}=\frac{1}{12\times 49}, when\ n=49$$

$$\frac{1}{m}=\frac{2}{12\times 50}=\frac{1}{6\times 50}, when\ n=50$$

When n = 53 or 55, m is not an integer. Hence, m and n take only 5 pairs of integral values.

18.The sum of 5 consecutive two digit even numbers, when divided by 10, becomes a perfect square. Which of the following cannot be one of those 5 numbers? (Number properties, Equations)

(A)76 (B)70 (C)54 (D)64 (E)36

Answer: option (D)

Let the five numbers are n, n+2, n+4, n+6, and n+8

Sum of these numbers = 5n+20

5n+20 = 40 or 90 or 160 or 250 or 360

Therefore, n = 4 or14 or28 or 46 or 68

The smallest and largest set of these five numbers is (4, 12) or (14, 22) or (28, 36) or (46, 54) or (68, 76).

It is clear from this that 64 does not fall into any of these sets.

19.A travels from X to Y at the speed of 60 mph. While A starts his journey from X at 3 PM, B starts her journey at 3.50 PM. They cross each other at 4 PM. The total journey-time of A and B together is 132 minutes and A reached Y before B reached X. Find the distance between X and Y. (Distance, Speed, Time)

A)110 miles B)132 miles C)72 miles D)82 miles E)92miles

Answer: option (C)

At 3.50 PM A has covered a distance of 50 miles. In another 10 minutes, he must have covered 10 miles. B took only 10 minutes to meet A. Let the distance traveled by B to meet A be d miles. Therefore, her speed is 6d.

At the time of crossing, their total journey-time is 60+10=70 minutes.

After crossing, they had traveled 132-70=62 minutes.

$$\frac{60}{6d}+\frac{d}{60}=\frac{62}{60}$$

Solving d=12 miles or 50 miles

When d=12, B would reach X at 4.50 PM and A would reach Y at 4.12 PM

When d=50, B would reach X at 4.12 PM and A would reach Y at 4.50 PM

Hence, the distance between X and Y is 72 miles.

20.Find the value of $\frac{1}{2!}+\frac{2}{3!}+\frac{3}{4!}+\frac{4}{5!}+\frac{5}{6!}+\frac{6}{7!}$(Series)

A)$\frac{7!-1}{7!}$B)$\frac{6!+1}{7!}$C)$\frac{6!-1}{7!}$D)$\frac{21}{7!}$E)$\frac{6}{7}$

Answer: option A

$$\frac{1}{2!}=\frac{2}{2!}-\frac{1}{2!}=1-\frac{1}{2!}$$

$$\frac{2}{3!}=\frac{3}{3!}-\frac{1}{3!}=\frac{1}{2!}-\frac{1}{3!}$$

Therefore, sum will be equal to

$$1 - \frac{1}{7!} = \frac{7! - 1}{7!}$$

21.A watermelon weighed 600 pounds. Ninety nine percent of its weight was due to the water content in the watermelon. After it was kept in a drying room for a while, it turned out that it was only 98% water by weight. How much does it weigh now? (Percentage, Proportions)

A)588 pounds B)594 pounds C)294 pounds D)300 pounds E)590 pounds

Answer: option D

99% of 600=594 pounds (water content); remaining 6 pounds is fiber.

After dried up, 6 pounds of fiber is 2%. We have to find 100%, that is, the present weight of the watermelon. When 2% corresponds to 6 pounds, 100% corresponds to 300 pounds (note that only water evaporates).

22.Nine teams play in a tournament. Each team plays every other team exactly once. A team receives 3 points for a win, 2 points for a draw, and 1 point for a loss. Every team ends up with a different total score and the difference between any two successive scores is same when they are ranked. The team with the highest total scored 20. What is the arithmetic mean score of the possible scores under loss that the team scored the lowest total? (Combinations, Progressions, and Average)

A)6 B)5 C)4 D)3 E)2

Answer: option B

9 teams play a total of 32 games ($9C_2$ ways)

The total scored by all the teams together is 32x4=144 (either 1 win and 1 loss, or 1 draw)

The nine scores are in arithmetic progression.

$$\frac{9}{2}(a + 20) = 144 => a = 12$$

There are 3 ways a team can score 12: (0 win, 4 draw, 4 loss) or (1 win, 2 draw, 5 loss) or (2 wins, 0 draw, 6 loss)

Therefore, the arithmetic mean of 4, 5, and 6 is 5

23.If the equation $(p+4)x^2-2px+2p-6=0$ is satisfied for all real values of x, the range of values of p is: (Equations)

A)[-6,4] B) (-6,4) C) [-6,4) D) (-6,4] E) $-6<p\leq4$

Answer: option A

If the equation $ax^2+bx+c=0$ is satisfied for all real values of x, then

$$b^2 - 4ac \geq 0$$

$$4p^2 - 4x(p + 4)(2p - 6) \geq 0$$

$$-4p^2 - 8p + 96 \geq 0$$

$$(4 - p)(p + 6) \geq 0$$

Therefore, $-6 \leq p \leq 4$

24.Find the ratio of total surface area of a cube to the largest tetrahedron, whose edges are equal, that could be kept inside the cube. (Solid geometry)

A)$2:\sqrt{3}$ B)$3:2\sqrt{3}$ C)$2\sqrt{3}:3$ D)$2\sqrt{3}:1$ E)$\sqrt{3}:1$

Answer: option E

Assume a unit cube

Total surface area of a unit cube is 6

Each edge of the tetrahedron will be equal to length of the diagonal of the side of the unit cube, which is equal to $\sqrt{2}$

The four triangles formed by the tetrahedron are equilateral.

Area of the triangles$=4 \times \frac{\sqrt{3}}{4}\sqrt{2}^2 = 2\sqrt{3}$

Therefore, the required ratio is $6:2\sqrt{3}$ or $\sqrt{3}:1$

25.What is the volume of the largest cube that can be kept inside a right circular cone whose radius and height are 1 unit and 3 units respectively? (Solid geometry, Similar triangle properties)

A)$\frac{36}{22+12\sqrt{2}}$ B)$\frac{108}{58+45\sqrt{2}}$ C)$\frac{54}{34+30\sqrt{2}}$ D)$\frac{36}{58+45\sqrt{2}}$ E)$\frac{54}{22+12\sqrt{2}}$

Answer: option B

The cube is at the center of the base of the cone. The cube's top 4 vertices touch the curved surface of the cone at some height. We have to find this height, which is the length of the edge of the cube. Draw a perpendicular line from the top vertex of the cone to the center of the base of the cone. This is the height of the cone, that is, 3 units. The center of the base of the cube and the cone is the same point. From the base center, draw a line through the diagonal of the cube, which touches the circumference of base of the cone. This line is the radius of the cone, that is, 1 unit. From the top center of the cube, draw a parallel line through the diagonal to the radius of the cone. Let the length of the edge of the cube be 'a' units. Apply similar triangle properties.

$$\frac{r}{R} = \frac{h}{H}$$

$$\frac{a/\sqrt{2}}{1} = \frac{h}{3}$$

$$h = \frac{3a}{\sqrt{2}}$$

$$a = 3 - \frac{3a}{\sqrt{2}}$$

$$a = \frac{6}{2 + 3\sqrt{2}}$$

$$a^3 = \frac{108}{58 + 45\sqrt{2}}$$

26.A right triangle has sides of length l, m, 27. Either l or m is the hypotenuse of the triangle. What is the arithmetic mean of all possible values of l and m? (Plane geometry and Average)

A)172.5 B)173.5 C)174.5 D)175.5 E)176.5

Answer: option D

We know that, $27^2 = l^2 - m^2 = (l + m)(l - m)$, assuming l>m

$$27^2 = 729 = 1 \times 729 = (365 - 364)(365 + 364) = 365^2 - 364^2$$

$$729 = 3 \times 243 = (123 - 120)(123 + 120) = 123^2 - 120^2$$

$$729 = 9 \times 81 = (45 - 36)(45 + 36) = 45^2 - 36^2$$

The possible values of l are 365, 123, and 45 and the possible values of m are 364, 120, and 36

The arithmetic mean of these values is 175.5

27.Suppose that you write 11 letters and then you address 11 envelopes to go with them. Closing your eyes, you randomly stuff one letter into each envelope. What is the probability that precisely two letters are in the wrong envelopes and all others in the correct envelopes? (Probability)

A)$\frac{1}{9!}$B)$\frac{1}{2\times 11!}$C)$\frac{1}{11!}$D)$\frac{1}{2\times 10!}$E)$\frac{1}{2\times 9!}$

Answer: option E

Select any two letters from the eleven. It can be done in $11c_2$ ways. The remaining 9 letters are in the correct envelopes.

The total number of ways that 11 letters can go into 11 envelopes is 11!

If you pick up one letter, that can go into any of the 11 envelopes.

The second letter can go into any of the remaining 10 envelopes, and so on.

The required probability is $11c_2/11!$

On simplification it is equal to 1/(2x9!)

28.Eight slips of paper with the numbers 1, 2, 3, 4, 5, 6, 7, and 8 written on them are placed into a bin. The eight slips are drawn one by one from the bin. What is the probability that the first four to come out are 1, 3, 5, 7 in some order? (Probability)

A)1/70 B)1/14 C)1/35 D)1/140 E)1/2

Answer: option A

Method 1

The first one can be drawn in $8c_1$ ways. It should be one of the four favorable numbers. Pick one from four in $4c_1$ ways.

The second one is done in $7c_1$ ways. It should be one of the remaining three favorable numbers. Pick one from the remaining three in $3c_1$ ways.

Similarly, the third and fourth slips can be drawn in $2c_1$ out of $6c_1$ and $1c_1$ out of $5c_1$ ways.

The required probability is (4/8)x(3/7)x(2/6)x(1/5)=1/70

Method 2

Pick any four slips from the eight slips in $8c_4$ ways.

The favorable numbers are 1, 3, 5, 7. Since order is not important, there is only one way it can be done.

The required probability is $1/8c_4$ or 1/70

29.A standard deck of 52 playing cards are divided into 4 sub-decks and kept on a table face down. What is the probability that one of those four top cards is a face card? (Probability)

A).396 B).553 C).496 D).453 E).353

Answer: option A

It is quicker if you apply 1 – P(not a face card)

There are 12 face cards and 40 other cards. Pick 4 cards from 40 other cards so that you do not get a face card. You can do that in $40c_4$ ways.

The total number of outcomes is $52c_4$

The required probability is $1 - (40c_4/52c_4) = .396$

30.Draw a planer grid that is 31 squares wide and 17 squares high. Then you remove squares so that the new grid is a 17 by 17-square grid. If a and b are the number of rectangles that can be drawn using the lines of the 31 by17 grid and 17 by 17 grid respectively then find (a-b).

A)52379 B)52479 C)51379 D)51479 E)50479

Answer: option B

In an m by n grid, the number of rectangles is $\frac{m(m+1)}{2} \times \frac{n(n+1)}{2}$

The same formula can be applied to an m by m - grid

Therefore, the required answer is $\frac{31\times32}{2} \times \frac{17\times18}{2} - \frac{17\times18}{2} \times \frac{17\times18}{2}$

On simplification, we get 52479

31.A pyramid with a slant height of 4 units in all its slant edges is cut at one corner of a cube with a side length of 8 units. Find the volume of the pyramid. (Solid geometry)

A)10.17 B)10.97 C)10.37 D)10.67 E)10.27

Answer: option D

Method 1

The pyramid has 1 equilateral triangle and 3 isosceles triangles. In the first method, we take the equilateral triangle as the base of the pyramid. The calculation is a bit more because you have to find the height of the pyramid.

Base area$=\frac{\sqrt{3}\times4\sqrt{2}\times4\sqrt{2}}{4}$ $=8\sqrt{3}$

You have to find the circum-radius of the equilateral triangle in order to find the height of the pyramid.

Area of triangle=abc/4R, where a, b, and c are sides of the triangle

$$R = \frac{4\sqrt{2}}{\sqrt{3}}$$

Height of the pyramid= $\sqrt{4^2 - (\frac{4\sqrt{2}}{\sqrt{3}})^2} = \frac{4\sqrt{3}}{3}$

Area of pyramid= $\frac{1}{3} \times h \times base\ area$

$$\frac{1}{3} \times \frac{4\sqrt{3}}{3} \times 8\sqrt{3} = \frac{32}{3} = 10.67$$

Method 2

It is easier and quicker. Take one of the isosceles triangles as the base of the pyramid. Now, height of the pyramid is 4 units. Base area= 8

Area of the pyramid $= \frac{1}{3} \times 4 \times 8 = \frac{32}{3} = 10.67$

32.Let B1 and B2 be two boxes such that B1 contains 3 white and 2 red balls and B2 contains only 1 white ball. A fair coin is tossed. If head appears then 1 ball is drawn at random from B1 and put into B2. However, if tail appears then 2 balls are drawn at random from B1 and put into B2. Now 1 ball is drawn at random from B2. Find the probability that the drawn ball from B2 being white. (Probability)

A)13/30 B)23/30 C)19/30 D)11/30 E)1/2

Answer: option B

When a fair coin is tossed, P(head)=1/2 and P(tail)=1/2

Case 1: head appears

The drawn ball from B1 is either white or red. If it is white then B2 contains 2 white balls. If it is red then B2 contains 1 white and 1 red.

P(white from B1)= $3c_1/5c_1$ and P(white from B2)=1 (since both are white)

P(red from B1)=$2c_1/5c_1$ and P(white from B2)=1/2

Therefore, P(white from B2)=$\frac{1}{2}\left\{\frac{3}{5}\times 1+\frac{2}{5}\times\frac{1}{2}\right\}=\frac{2}{5}$

Case 2: tail appears

The two balls drawn from B1 can be 2w or 1w and 1r or 2r

P(2w from B1)=$3c_2/5c_2$ and P(w from B2)=1 (since all are white)

P(1w and 1r from B1)=($3c_1$x$2c_1$)/$5c_2$ and P(w from B2)=$2c_1/3c_1$

P(2r from B1)=$2c_2/5c_2$ and P(w from B2)=$1c_1/3c_1$

Therefore, P(white from B2)=$\frac{1}{2}\left\{\frac{3}{10}\times 1+\frac{3}{5}\times\frac{2}{3}+\frac{1}{10}\times\frac{1}{3}\right\}=\frac{11}{30}$

Hence, the required probability is $\frac{2}{5}+\frac{11}{30}=\frac{23}{30}$

33.A and B throw 2 dice at a time, one by one. If sum is 8 then thrower wins. A starts the game and if probability of his winning is x/(2x-5) find the value of x. (Probability, Progressions)

A)36 B)67 C)-36 D)-67 E)-25/26

Answer: option A

There are five favorable cases: (2,6), (3,5), (4,4), (6,2), (5,3)

P(success)=5/36 (total outcome is 36)

P(failure)=31/36

Probability of A's win= P(s)+P(f).P(f).P(s)+P(f).P(f).P(f).P(f).P(s)+......

$$\frac{5}{36}+\frac{31}{36}\times\frac{31}{36}\times\frac{5}{36}+\frac{31}{36}\times\frac{31}{36}\times\frac{31}{36}\times\frac{31}{36}\times\frac{5}{36}+\cdots$$

It is a geometric progression and the series in infinite, where a=5/36 and r=$(31/36)^2$

$$S_\infty = \frac{a}{1-r}$$

$$\frac{5/36}{1-\frac{961}{1296}} = \frac{36}{67}$$

$$\frac{x}{2x-5} = \frac{36}{67}$$

$$=> x = 36$$

34.A two digit number is 6 more than another two digit number. Then the first number is written in base 5 and the second number is written in base 4. It happened that the second number is the reverse of the first number. Find the product of these two numbers in base 10. (Number properties)

A)156 B)180 C)165 D)187 E)209

Answer: option D

Let the number in base 5 be xy and the number in base 4 be yx

$$y \times 5^0 + x \times 5^1 = x \times 4^0 + y \times 4^1 + 6$$

$$y + 5x = x + 4y + 6$$

$$3y = 4x - 6$$

'x' cannot take the value of 0 and 9. It assumes only one value, that is, 3.

When x=3, y=2

Therefore, the number in base 5 is 32 and the number in base 4 is 23.

$$(32)_5 = (17)_{10}$$

$$(23)_4 = (11)_{10}$$

Therefore, the product of the two numbers is 187.

35.B&W Cargo charges $16000 as the freight charges when the distance to be covered is 1000 miles and $10000 for a distance of 550 miles. Find the freight charge for 280 miles. (Ratio and proportions)

A)$5200 B)$5100 C)$5900 D)$6100 E)$6400

Answer: option E

When the distance is reduced by 450 miles, the freight charge is reduced by $6000. Therefore, when the distance is reduced by 720 miles, the freight charge will be reduced by $9600.

$$\frac{6000}{450} \times 720 = 9600$$

Therefore, the freight charge for 280 miles is 16000-9600=$6400

36.An empty tank is half filled with milk. Then it is filled with water up to half of the remaining. Then it is filled with milk up to half of the remaining. The process continues until the tank is filled. Find the ratio of milk to water. (Ratio, Progressions)

A)1:1 B)3:1 C)2:1 D)4:1 E)3:2

Answer: option C

Let us calculate the milk part first.

$$\frac{1}{2} + \frac{1}{8} + \frac{1}{32} + \cdots$$

This is a geometric progression whose terms are infinite, where a=1/2, r=1/4

$$S = \frac{a}{1-r} = \frac{1/2}{1-1/4} = \frac{2}{3}$$

It means milk is two-third of the tank and water is one-third of the tank.

The required ratio is 2:1

37.A can do a piece of work in 32 days. A and B together can do the same piece of work in 8 days. They undertook a work, which they can complete in 40 days, if they worked together, for $16000. For the first 10 days, B was absent and then they completed the remaining part working together. In how many days did they complete the whole work? (Work and Time)

A)50 days B)47.5 days C)52.5 days D)52 days E)42.5 days

Answer: option B

A and B together can complete a piece of work in 8 days. Therefore, in 40 days they can complete 5 times of the work.

'A' can do in one day 1/32 work; in 10 days he can do 10/32 or 5/16 work.

The remaining work is 5-(5/16)=75/16 work

A and B together can complete in one day 1/8 work

Therefore, they take $\frac{75/16}{1/8} = \frac{75}{2} = 37.5\ days$

Hence, they complete the whole task in 47.5 days.

Method 2

They together can complete the task in 40 days. That means A alone will take 160 days to complete the task. In 10 days, A will complete 1/16 work.

The remaining work of 15/16 is completed by A and B together in 37.5 days.

38.In the above question, what is the ratio of the remuneration of A to B?

A)10:33 B)19:45 C)17:45 D)1:3 E)2:5

Answer: option B

If A can do a work in x days and along with B can do the same work in y days, then efficiency of A to B is y:(x-y).

Therefore, in the above question efficiency of A to B is 1:3.

Hence, A and B share their remuneration in this ratio.

The contract amount is $16000 for 40 days. In one day, they earn $400.

Therefore, A earns $100 per day and B earns $300 per day according to their efficiency ratio.

A works for 47.5 days and his remuneration is $4750.

B earns the remaining amount of $11250.

The required ratio is 19:45

39.A man bought two tables for $1440. He sold one of them at a gain of 15% and the other at a loss of 8%. It was then found that each table fetched him the same amount. Find the cost price of second table. (Profit and Loss, Ratio)

A)$800 B)$740 C)$940 D)$640 E)$880

Answer: option A

Let the cost price of the first table be x and the second table be y.

$$1.15x = .92y$$

$$=> \frac{x}{y} = \frac{92}{115}$$

The ratio of cost price of the first table to second table is found.

Total cost price is $1440.

$$(92 + 115) \rightarrow \$1440$$

$$115 \rightarrow \frac{1440}{207} \times 115 = \$800$$

40.If x travels at 30 miles per hour, then x reaches the destination late by 10 minutes. If x travels at 42 miles per hour, then x reaches the destination 10 minutes early. Find the distance. (Time, Speed, and Distance)

A)30 miles B)42 miles C)38 miles D)35 miles E)40 miles

Answer: option D

Method 1

Distance is constant. Let t be the usual time.

$$30\left(t + \frac{1}{6}\right) = 42(t - \frac{1}{6})$$

$$\frac{6t + 1}{6t - 1} = \frac{42}{30}$$

$$=> t = 1\ hour$$

Therefore, distance=35 miles

Method 2

If x travels at s_1 mph and takes t_1 time more than usual time and if x travels at s_2 mph and takes t_2 time less than the usual time to reach a destination, then usual time t:

$$t = \frac{s1t1 + s2t2}{s2 - s1}$$

Once you find the usual time you can calculate distance.

41.X, Y, and Z are three typists who working simultaneously can type 216 pages in four hours. In one hour X can type as many pages more than Y as Y can type more than Z. During a period of five hours, X can type as many pages as Z can during seven hours. How many pages does Y type per hour? (Work, Ratio)

A)15 B)16 C)17 D)18 E)19

Answer: option D

Time ratio of X to Z is 5:7; therefore, efficiency ratio of X to Z is 7:5

Let X types 7p pages in one hour and Z types 5p pages in one hour.

From the second sentence, we conclude that Y types 6p pages in one hour.

From the first sentence, they type 54 pages in one hour.

Hence, 18p=54; p=3

Y types 18 pages in one hour.

42.A can do a piece of work in 18 days, B in 9 days and C in 6 days. A and B start working together and after 2 days C joins them. What is the total number of days taken to finish the work? (Work, Time)

A)6 B)5 C)4 D)3.5 E)3

Answer: option C

If A takes x days and B takes y days to complete a similar work, then A and B together will take $\frac{x \times y}{x+y}\ days$to complete the same work.

A and B together can complete the work in 6 days. In 2 days, they complete one-third of the work. A, B, and C together can complete the work in 3 days. They have to complete two-third of the work. Therefore, they take 2 days to complete the remaining work. Hence, the total number of days to complete the work is 4.

43.A is twice as good a workman as B and finished a piece of work in t hours less than B. In how many hours they together could finish that piece of work. (Work, Time)

A)3t/2 B)2t/3 C)2t D)3t E)4t/3

Answer: option B

Let B takes x hours to complete that piece of work, then A takes x-t hours.

Also, x=2(x-t)

It implies x=2t hours and A=t hours.

$$\frac{t \times 2t}{t + 2t} = \frac{2t}{3}$$

44.From a barrel containing 500 liters of milk, 3 mugs of milk are poured into a barrel containing 1000 liters of water. After mixing the contents well, 3 mugs of the mixture are poured into the barrel of milk. What is the ratio of the percentage of water in the barrel of milk to the percentage of milk in the barrel of water? (Mixture, Ratio, Percentage)

A)1:1 B)2:1 C)1:2 D)2:3 E)3:2

Answer: option B

Note that given a certain ratio of the initial quantity of liquids, say a:b, in the two barrels, if certain quantity of the first liquid is poured into the second and the same quantity is taken from the second and poured into the first, the ratio of percentage of second liquid in the first barrel to percentage of first liquid in the second barrel will always be b:a, that is, the inverse ratio.

Therefore, we can assume any quantity say, 100 liters of milk is poured into the second barrel.

Now, the second barrel contains 100 liters of milk and 1000 liters of water. The ratio is 1:10.

If you take 100 liters of the mixture, it will contain 1/11 part of milk and 10/11 part of water.

After pouring the mixture in the first barrel, water in the first barrel will be 1000/11 liters.

Milk in the second barrel will be 1000/11 liters.

Ratio of percentage of water in the first to percentage of milk in the second is

$$\frac{1000/11}{500}:\frac{1000/11}{1000}$$

Solving, we get 2:1

45.An alloy contains copper and zinc in the ratio 5:3 and another allow contains copper and tin in the ratio 8:5. If equal quantities of both the alloys are melted together to form another alloy then what is the percentage of tin in the resultant alloy. (Mixture, Percentage)

A)500/21 B)200/13 C)150/13 D)175/13 E)250/13

Answer: option E

If you take 1 kg of the first alloy $(\frac{5}{8}+\frac{3}{8})$ and 1 kg of the second alloy $(\frac{8}{13}+\frac{5}{13})$, then tin in the resultant alloy will be 5/13 out of 2 kg. To find percentage multiply with 100.

$$\frac{5}{13}\times\frac{1}{2}\times 100=\frac{250}{13}$$

46.If $1066 are divided among A,B,C, and D such that A:B=3:4; B:C=5:6; and C:D=7:5. Who will get the maximum? (Ratio and Proportions)

A)A B)B C)C D)D E)E

Answer: option C

B gets more than A; C gets more than B; therefore, C>B>A.

Between C and D, C gets more than D. Hence, C gets the maximum.

47.The total emoluments of A and B are equal. However, A gets 65% of his basic salary as allowances and B gets 80% of his basic salary as allowances. What is the ratio of the basic salaries of A and B? (Ratio, Percentage)

A)11:12 B)16:13 C)12:11 D)13:16 E)13:12

Answer: option C

Let their gross salaries be GA and GB; GA=GB.

GA=basic salary of A+.65 of basic salary of A=1.65 of basic salary of A

GB=basic salary of B+.8 of basic salary of B=1.8 of basic salary of B

$$\frac{basic\ salary\ of\ A}{basic\ salary\ of\ B} = \frac{1.8}{1.65} = \frac{12}{11}$$

48.A family wanted to reduce the expenditure on milk by 19%. However, the price of milk increased by 8%. In order to achieve the desired result, by what percentage the family would reduce their consumption? (Percentage)

A)20% B)27% C)26.2% D)23.75% E)25%

Answer: option E

Method 1

Let the family buy 100 liters of milk for $100. The family now has only $81 to buy milk. However, 100 liters of milk costs $108, after increase.

Number of liters the family can buy=81/1.08 = 75 liters.

Therefore, the family has to reduce their consumption by 25%.

Method 2

If r is equal to the product of two other quantities, say p and q, then the percentage change in r with respect to the percentage changes in p and q can be obtained by a formula:

$$\textit{The percentage change in } r = x + y + \frac{x \times y}{100}, \textit{where } x \textit{ and } y \textit{ are \% changes in } p \textit{ and } q.$$

In this problem, the % change in r is given as -19 (negative indicates decrease). The % change in milk is 8 (positive indicates increase). We have to find the % change in quantity (negative).

$$8 - q - \frac{8 \times q}{100} = -19 => q = 25$$

49.A and B run a 1400 meter race. In the first heat, A gives B a start of 35m and beats B by 40 seconds. In the second heat, A gives B a start of 60 seconds and is beaten by 40 meter. In what time could A run a 2 km race? (Race problems)

A)500 sec B)400 sec C)450 sec D)480 sec E)350 sec

Answer: option A

Let A's speed be s_1 and B's speed be s_2.

In the first heat, A beats B by 40 seconds. It implies A beats B by $40s_2$ meter. When A covers 1400 m, B covers a distance of $1400-35-40s_2$ m, that is, $1365-40s_2$ meter (1).

In the second heat, when A covers 1360 m B covers a distance of $1400-60s_2$ m. Therefore, when A covers 1400 m B will cover (2)

$$(1400 - 60s_2)\frac{1400}{1360}$$

Equating (1) and (2), we get

$$1365 - 40s_2 = (1400 - 60s_2)\frac{35}{34}$$

Solving, s_2=3.5 m/sec.

In the first heat, when A covered 1400 m B covered only 1400-35-140=1225m and he took (1225/3.5) = 350 seconds. This is the time taken by A to cover 1400 m and A's speed is (1400/350) = 4 m/sec.

That means A will take (2000/4) = 500 seconds to cover a 2 km race.

50.In what scale notation is 25 doubled by reversing the digits.

A)5 B)6 C)7 D)8 E)9

Answer: option D

Method 1

Let the radix be x.

Therefore, $25=5+2.x^1=5+2x$; $52=2+5.x^1=2+5x$

By supposition, $2+5x=2(5+2x)$; $x=8$

Method 2

The radix is greater than 5. Take the middle value in the option.

5+2x7=19; 2+5x7=37. Thirty seven is less than two times of 19. Therefore, we have to increase the radix. Take 8.

5+2x8=21; 2+5x8=42. It fits the supposition.

51.A waterman rows to a place 48 miles distance and back in 14 hours: he finds that he can row 4 miles with the stream in the same time as 3 miles against the stream. Find the rate of the stream. (Time, Speed, Distance)

A)7 mph B)5 mph C)3 mph D)2 mph E)1 mph

Answer: option E

Let the speed of the waterman with the stream and against the stream be x and y mph respectively.

$$\frac{48}{x}+\frac{48}{y}=14$$

$$\frac{x}{y}=\frac{4}{3}=>x=\frac{4y}{3}$$

Substituting, we get x=8 and y=6.

Therefore, speed of the stream is $\frac{8-6}{2}=1$ mph.

52.A family consumes 20 loaves of bread in a week. If wages were raised 5 percent and the price of bread were raised 2.5%, the family would gain 30 cents a week. However, if wages were lowered 7.5% and bread fell 10%, then the family would lose 10 cents a week. Find the price of a loaf. (Percentage, Equations)

A)32 cents B)30 cents C)28 cents D)26 cents E)24 cents

Answer: option C

Let w, p, and s be wages, cost of bread, and usual savings respectively.

Wages-cost of bread=usual savings.

By supposition,

1.05w-1.025p=s+30 cents .....(1)

.925w-.9p=s-10cents ....(2)

Solving, we get w-p=$3.20 (subtracting 2 from 1)

As w-p=s, substitute s=3.20 in equation (1) and (2)

Solving again, we get the price of a loaf as 28 cents (Hint: add the equations and substitute the value of w-p).

53.A train, an hour after starting, meets with an accident which detains it an hour, after which it proceeds at three-fifths of its former speed and arrives 3 hours late. However, had the accident happened 100 km farther on the line, it would have arrived 1.5 hours sooner. Find the length of the journey. (Time, Speed, Distance)

A)1600/9 km B)1500/9 km C)1400/9 km D)1300/9 km E)1200/9 km

Answer: option A

Let the usual speed of the train be s km per hour and usual time t hours.

It meets with an accident after 1 hour. It means it covered a distance of s km. It should take (t-1) hours to cover the remaining distance with usual speed. However, it takes (t+1) hours to cover the remaining distance with three-fifths of its original speed (t+3-1-1). Since time and speed are inversely proportional, it will take $\frac{5}{3}(t-1)$ hours.

Therefore, $\frac{5}{3}(t-1) = t + 1 \Rightarrow t = 4\ hours.$

Had the accident happened 100 km farther, it would have arrived 1.5 hours sooner. This hundred kilometers distance makes the difference.

$$\frac{100}{3s/5} - \frac{100}{s} = 1.5$$

Solving, we get s=400/9. Total distance, 4s=1600/9 km.

54.In what time A,B, and C together do a work if A alone could do it in six hours more, B alone in one hour more, and C alone in twice the time? (Work and Time, Quadratic equations)

A)70 min B)60 min C)50 min D)40 min E)30 min

Answer: option D

Let A, B, and C together can complete the work in x days.

It is given that C alone can do the same work in 2x days. It implies C would complete half of that work in x days. Half of the remaining work is completed by A and B together.

$$\frac{(x+6)(x+1)}{2x+7} = 2x$$

Solving, we get x=2/3 hours or 40 minutes.

55.A vessel contains 15 liters of wine and another vessel contains 10 liters of water: 6 liters are taken out of each vessel and transferred to the other; this operation is repeated for 4 times. Find the percentage of wine in the first vessel after the fourth operation. (Mixture, Ratio, and Percentage)

A)20% B)30% C)40% D)50% E)60%

Answer: option E

In the first operation, 6 liters of wine and 6 liters of water are taken out and transferred to the other. Therefore, the first vessel now contains 9 liters of wine and 6 liters of water and the second vessel 4 liters of water and 6 liters of wine.

Look at the ratio of wine to water in the first vessel and the second vessel. The ratio is same in both the vessels. Therefore, the ratio will not change after further operations. The ratio will remain constant after the first operation.

Therefore, the percentage of wine in the first vessel is 60%.

56.In a certain community consisting of p persons, a percent can read and write; of the males alone b percent, and of the females alone c percent can read and write. Find the number of females in the community if p=900, (a-c)=10, and (b-c)=15. (Percentage, Equations)

A)300 B)400 C)500 D)600 E)200

Answer: option A

Let number of males is x and number of females p-x.

Read and write (total) $=\frac{pa}{100}$

Read and write (males) $=\frac{xb}{100}$

Read and write (females) $=\frac{(p-x)c}{100}$

By supposition, $pa = xb + (p - x)c$

$$pa = xb + pc - xc$$

$$x = \frac{p(a-c)}{(b-c)}$$

$$x = \frac{900 \times 10}{15} = 600$$

$$p - x = 300$$

57.The interior angles of a pentagon are in a ratio of an arithmetic progression whose first term is 4 and common difference is 1. Find the difference between the largest and the smallest angle. (Geometry, Ratio, Progressions)

A)108 B)71 C)54 D)72 E)60

Answer: option D

Let the ratio be 4:5:6:7:8.

Let the angles be 4x, 5x, 6x, 7x, and 8x.

Sum of interior angles of any polygon=$(n-2)180, \textit{where n is number of sides}$

By supposition, $4x + 5x + 6x + 7x + 8x = (5-2)180.$

$$30x = 3 \times 180 => x = 18$$

$$8x - 4x = 4x = 4 \times 18 = 72$$

58.The complement of an angle exceeds the angle itself by 40°. Find the angle.

A)70° B)65° C)60° D)55° E)50°

Answer: option B

If the angle is x, then the complement angle is 90-x.

By supposition, $x - (90 - x) = 40 => x = 65°$

59.The supplement of an angle is one-fifth of the angle itself. Find the angle.

A)130° B)140° C)150° D)160° E)120°

Answer: option C

Let the angle be x and the supplement angle be $\frac{x}{5}$.

By supposition, $x + \frac{x}{5} = 180 => x = 150°$

60.In a triangle ABC, $AB^2+AC^2=200$ cm^2. Median, AD=8 cm. find BC. (Properties of triangles)

A)16 B)15 C)10√2 D)9√2 E)12

Answer: option E

We know that, $AB^2+AC^2=2(AD^2+BD^2)$

$$200 = 2(64 + BD^2) => BD^2 = 36; BD = 6$$

Therefore, BC=12 cm.

61.The exterior angle of a regular polygon is one-third of its interior angle. How many sides has the polygon?

A)9 B)8 C)7 D)6 E)5

Answer: option B

$$\text{The exterior angle of any regular polygon} = \frac{360°}{n}, \text{where } n \text{ is the number of sides}$$

$$\text{The interior angle of any regular polygon} = \frac{(n-2)}{n} \times 180°$$

By supposition,

$$\frac{360}{n} = \frac{1}{3} \times \frac{(n-2)}{n} \times 180$$

Solving, we get n=8.

62.Two regular polygons are such that the ratio between their number of sides is 1:2 and the ratio of measures of their interior angles is 3:4. Find the interior angle of the first polygon.

A)140° B)135° C)120° D)108° E)90°

Answer: option D

Let the number of sides of the two polygons be x and 2x respectively.

By supposition,

$$\frac{(x-2)}{x} \times 180 : \frac{(2x-2)}{2x} \times 180 :: 3:4$$

Solving, x=5 and 2x=10.

The interior angle of the first polygon is,

$$\frac{5-2}{5} \times 180 = 108°$$

63.One of the angles of a triangle is 144°. Find the angle between the bisectors of the acute angles of the triangle.

A)162° B)144° C)126° D)118° E)108°

Answer: option A

Let ABC be the triangle and angle B be 144°.

Sum of the acute angles of triangle ABC is 36°.

Angle between the bisectors of the acute angles= $180 - \frac{(A+C)}{2} = 180 - 18 = 162°$

64.D and E are the middle points of the sides AB and AC of a triangle ABC. Find the ratio of the area of triangle ADE to area of triangle ABC.

A)1:2 B)1:3 C)1:4 D)2:3 E)3:5

Answer: option C

Since D and E are the mid-points, triangle ADE and triangle ABC are similar. The sides of triangle ADE and ABC are in the ratio 1:2. Therefore, the ratio of area of triangle ADE to area of triangle ABC is 1:4.

$$\frac{1}{2} \times DE \times h : \frac{1}{2} \times 2DE \times 2h$$

$$1:4$$

65.In a triangle ABC, the base angles B and C are 45° each. D is mid-point on BC. AD=50 cm. Find the length of BC.

A)$50\sqrt{2}$ cm B)$50/\sqrt{2}$ cm C)50 cm D)$100/\sqrt{3}$ cm E)100 cm

Answer: option E

AD is a median. BD=DC. AD is also altitude drawn from A to BC (AB=AC). Therefore, DC=AD (angle BCA=angle CAD=45°). Hence, BC=100 cm.

66.ABC is right angled triangle at A. Angle B is twice angle C. Which of the following is true?

A)$AC^2=2AB^2$ B)$2AC^2=AB^2$ C)$3AC^2=AB^2$ D)$AC^2=3AB^2$ E)$AC^2=4AB^2$

Answer: option D

Property of 30-60-90 triangle states sides opposite of angles are in the ratio $1:\sqrt{3}:2$. Angle B is 60° and angle C is 30°. If side opposite of angle B is $\sqrt{3}$, then side opposite of angle C is 1. Therefore, $AC^2=3AB^2$.

67.The present value of $672 due in a certain time is $126. If compounded interest at $4\frac{1}{6}\%$ be allowed, find the time.

A)17 years B)25 years C)35 years D)41 years E)53 years

Answer: option D

$126 doubles in approximately in $\frac{72}{4\frac{1}{6}}$ years, that is, approximately in 17 years and 4 months.

$126 becomes four times in approximately another 17 years and 4 months.

Now, you can eliminate options A, B, and C.

$126 becomes eight times in approximately another 17 years and 4 months.

Option E is also eliminated.

Note: if you want to apply compound interest formula you need to know the values of log2 and log3 which are .3010 and .4771 respectively. The calculation is quite lengthy.

68.At simple interest the interest on a certain sum of money is $90, and the discount on the same sum for the same time and at the same rate is $80. Find the sum.

A)$640 B)$660 C)$680 D)$700 E)$720

Answer: option E

Let p be the sum and the present value of p be v. Therefore,

$$\frac{pnr}{100} = 90$$

$$discount = p - v = 80$$

$$p = v + \frac{vnr}{100} => \frac{vnr}{100} = 80$$

$$\frac{p}{v} = \frac{90}{80}$$

$$\frac{p}{p-80} = \frac{9}{8} => p = \$720$$

69.A freehold estate is bought for $275,000; at what rent per annum should be let so that the owner may receive 4 per cent per annum on the purchase money allowing compound interest?

A)$11,000 B)$10,500 C)$10,000 D)$9,500 E)$9,000

Answer: option A

Let A be the annual rent and V be the present values of all annual rents received.

The present value of the first rent=$A(1.04)^{-1}$

The present value of the second rent=$A(1.04)^{-2}$

Therefore, the present values of all rents,

$$V = A(1.04)^{-1} + A(1.04)^{-2} + A(1.04)^{-3} + \cdots$$

$$V = \frac{A}{0.04}$$

(RHS is geometric series, where r is $(1.04)^{-1}$ and number of terms is infinite.)

$$275000 = \frac{A}{0.04}$$

$$A = \$11{,}000$$

Note: If you know, $V = \frac{A}{r}$ then it is a one-liner.

70.Three travelers arrive at a town where there are four hotels; in how many ways can they take up their quarters, each at a different hotel?

A)6 B)12 C)16 D)18 E)24

Answer: option E

The first person can choose anyone of the four hotels. The second person can choose anyone of the remaining three hotels. The third person can choose anyone of the remaining two hotels. Therefore, they can take up the quarters in 24 ways.

71.In how many ways can 15 recruits be divided into three equal groups?

A)126126 B)252252 C)378378 D)504504 E)756756

Answer: option A

The number of ways p+q+r things can be divided into three groups containing p, q, and r things respectively is $\frac{(p+q+r)!}{p!q!r!}$

In this problem, p+q+r=15; p=q=r. We have to modify the formula. If the groups have equal number of things then we use

$$\frac{3p!}{p!\,p!\,p!\,3!}$$

$$\frac{15!}{5!\,5!\,5!\,3!} = 126126$$

72.The duration of a railway journey varies directly as the distance and inversely as the velocity; the velocity varies directly as square root of the quantity of coal used per mile, and inversely as the number of carriages in the train. In a journey of 50 miles in half an hour with 18 carriages 100 kg of coal is consumed; how much coal will be consumed in a journey of 42 miles in 28 minutes with 16 carriages? (Variation)

A)64.76 B)80.76 C)53.76 D)52.76 E)51.76

Answer: option C

$$t \propto \frac{d}{v} \ and \ v \propto \frac{\sqrt{q}}{c}$$

$$t \propto \frac{dc}{\sqrt{q}} => t = \frac{kdc}{\sqrt{q}}$$

By supposition,

$$\frac{1}{2} = k \times \frac{50 \times 18}{\sqrt{2}} => k = \frac{\sqrt{2}}{1800}$$

$$\therefore \frac{28}{60} = \frac{\sqrt{2}}{1800} \times \frac{42 \times 16}{\sqrt{q}} => q = \frac{32}{25} \ kg \ per \ mile$$

Total consumption of coal is $\frac{32}{25} \times 42 = 53.76\,kg.$

73.A man starts from A to B, another starts from B to A at the same time. After they meet, they complete their journeys in 10/7 and 14/5 hours respectively. Find the speed of the second man if the speed of first is 70 miles per hour. (Time, Speed, and Distance)

A)65 mph B)60 mph C)55 mph D)50 mph E)45 mph

Answer: option D

Method 1

Let t hours be the time taken by A and B when they meet.

Therefore, A covered 70t miles when he met B and thereafter A covered 100 miles to complete the journey.

B took 14/5 hours to cover 70t miles. Therefore, B's speed is $\frac{70t}{14/5} = 25t$

By supposition,

$$\frac{100}{25t} = t => t = 2\ hours$$

Therefore, B's speed is 50 miles per hour

Method 2

$$\frac{A's\ speed}{B's\ speed} = \sqrt{\frac{B's\ time\ taken\ to\ complete\ the\ remaining\ journey}{A's\ time\ taken\ to\ complete\ the\ remaining\ journey}}$$

$$\frac{70}{B's\ speed} = \sqrt{\frac{14/5}{10/7}} = \frac{7}{5}$$

Therefore, B's speed is 50 miles per hour.

74.I shall be 40 minutes late to reach my office if I walk from my house at 3 miles per hour. I shall be 30 minutes early if I walk at 4 miles per hour. Find the distance between my house and the office.

A)10 miles B)11 miles C)12 miles D)13 miles E)14 miles

Answer: option E

Method 1

Let the distance be d, usual time taken be t, and usual speed be s.

By supposition,

$$\frac{d}{3} = t + \frac{2}{3} - - - eqn\ 1$$

$$\frac{d}{4} = t - \frac{1}{2} - - - eqn\ 2$$

$$\frac{d}{12} = \frac{7}{6}(eqn\ 1 - eqn\ 2)$$

Therefore, distance is 14 miles.

Method 2

Let us take a convenient number for the distance. The convenient number is 12 miles (LCM of 3 and 4).

At 3 miles per hour, the time taken is 4 hours; and at 4 miles per hour, the time taken is 3 hours. The difference is 60 minutes. However, in the problem, the difference is 70 minutes (40 min+30 min).

60 min → 12 miles

70 min → 14 miles

75.A and B walk from P to Q a distance of 21 miles at 3 and 4 miles per hour respectively. B reaches Q and immediately returns and meets A at R. Find the distance from P to R.

A)14 miles B)15 miles C)16 miles D)17 miles E)18 miles

Answer: option E

Together A and B cover a distance of 42 miles. A covers the distance from P to R and B covers the distance from P to Q and then from Q to R. The ratio of the speeds of A and B is 3:4. Therefore, the distances covered by them will also be in the same ratio. Hence, the distance from P to R is $\frac{3}{7} \times 42 = 18\, miles.$

76.A train after travelling 50 miles from A meets with an accident and proceeds at four-fifths of the former speed and reaches B, 45 minutes late. Had the accident happened 20 miles further on, it would have arrived 12 minutes sooner. Find the distance between A and B.

A)100 miles B)125 miles C)140 miles D)150 miles E)175 miles

Answer: option B

In the first case, the accident took place at 50 miles; in the second case, the accident took place at 70 miles. In the first case, the next 20 miles were covered at four-fifths of original speed; in the second case, it was covered at the usual speed. By supposition,

$$\frac{20}{4s/5} - \frac{20}{s} = \frac{12}{60} => s = 25\,miles\;per\;hour$$

In the first case, let x be the remaining distance to be covered after the accident. By supposition,

$$\frac{x}{20} - \frac{x}{25} = \frac{45}{60} => x = 75\,miles$$

Therefore, the total distance from A to B is 125 miles.

77.A train leaving L at 3.10 p.m. reaches W at 5.00 p.m. One leaving W at 3.30 p.m. arrives in L at 5.50 p.m. When do they pass each other?

A)16:18:24 B)16:19:24 C)16:20:24 D)16:21:24 E)16:22:24

Answer: option C

The train that leaves L takes 110 minutes to reach W. The train that leaves W takes 140 minutes to reach L. The ratio of time taken is 11:14. Therefore, the ratio of their speeds will be 14:11. Let the speeds of the first and second train be 14s and 11s respectively. Therefore, the distance between L and W will be:

$$14s \times \frac{110}{60} = \frac{77s}{3}$$

At 3.30 p.m., the first train must have covered a distance of:

$$\frac{20}{60} \times 14s = \frac{14s}{3}$$

Therefore, the time taken to meet is:

$$\frac{63s/_3}{14s + 11s} = \frac{21}{25} \ hour\ after\ 3.30\ p.m.$$

78.If $x<1$, find the sum of the series $1+2x+3x^2+4x^3+....\infty$

A)$\frac{1}{1-x}$ B)$\frac{1}{(1-x)^2}$ C)$\frac{x}{1-x}$ D)$\frac{x}{(1-x)^2}$ E)$\frac{1-x}{x}$

Answer: B

If you observe the series, each term is a product of corresponding terms in an arithmetic and geometric series.

1+2+3+4+.... is an arithmetic series and $x^0+x^1+x^2+x^3+....$ is a geometric series.

In the arithmetic series, a=1 and d=1.

In the geometric series r=x.

The sum of infinite terms of such a series is:

$$\frac{a}{1-r} + \frac{dr}{(1-r)^2}$$

$$\therefore \frac{1}{1-x} + \frac{x}{(1-x)^2}$$

$$= \frac{1-x+x}{(1-x)^2}$$

$$= \frac{1}{(1-x)^2}$$

79.Find the sum to infinity of the series:

$$1 + \frac{4}{5} + \frac{7}{5^2} + \frac{10}{5^3} + \cdots$$

A)16/35 B)17/35 C)35/17 D)35/16 E)35/18

Answer: D

The given series is product of arithmetic and geometric series, where a=1, d=3, and r=1/5. Sum to infinity of such series is:

$$\frac{a}{1-r} + \frac{dr}{(1-r)^2}$$

$$\therefore \frac{1}{1-{}^1/_5} + \frac{3 \times {}^1/_5}{(1-\frac{1}{5})^2}$$

$$\frac{5}{4} + \frac{3}{5} \times \frac{25}{16}$$

$$= \frac{35}{16}$$

80.The arithmetic and geometric mean of two numbers are 6.5 and 6 respectively. Find the harmonic mean of the two numbers.

A)5.938 B)5.838 C)5.738 D)5.638 E)5.538

Answer: E

Arithmetic mean x Harmonic mean = (Geometric mean)^2

$$6.5 \times HM = 36 => HM = \frac{36}{6.5} = 5.538$$

81.Find the sum of the series, 1x2+2x3+3x4+...., to 20 terms.

A)1540 B)3080 C)2540 D)1580 E)3040

Answer: B

The nth term of the series is n(n+1).

We have to find $\Sigma n(n+1)=\Sigma n^2+\Sigma n=\frac{n(n+1)(n+2)}{3}$, where n=20.

$$\sum n^2 = \frac{n(n+1)(2n+1)}{6}$$

$$\sum n = \frac{n(n+1)}{2}$$

$$\therefore \frac{20 \times 21 \times 22}{3} = 3080$$

82.Find the minimum value of x^2-4x+7 for real values of x.

A)2 B)-2 C)3 D)-3 E)1

Answer: C

$$x^2 - 4x + 7 = (x-2)^2 + 3$$

$$since\ (x-2)^2\ is\ positive,$$

$$the\ value\ of\ the\ expression\ is\ minimum$$

$$when\ (x-2)^2\ is\ zero. That\ is, when\ x = 2.$$

Therefore, the minimum value of the expression is 3.

83.If x is positive, find the greatest value of (5-x)(x+3).

A)20 B)19 C)18 D)17 E)16

Answer: E

$$(5-x)(x+3) = 15 + 2x - x^2$$

$$= 16 - (1-x)^2$$

The given expression is maximum when $(1-x)^2$ is zero. The maximum value of the expression is 16 when x=1.

84.Find the solution set of x if |2x-5|<4x+9

A)x<2/3 B)x<-2/3 C)x>2/3 D)x>-2/3 E)x>-7

Answer: D

2x-5<4x+9 or -2x+5<4x+9

Case:1

$2x-5<4x+9$

$-2x<14$ or $2x>-14$

$x>-7$

Case:2

$-2x+5<4x+9$

$-6x<4$ or $6x>-4$

$x>-2/3$

combining the results of case 1 and case 2, we get $x>-2/3$.

85.Find the area of the quadrilateral formed by the solution set of the inequalities $2x+3y\leq12$, $x\geq0$, $y\geq0$ and $x\leq3$.

A)9 B)8 C)7 D)6 E)5

Answer: A

The x-intercept and y-intercept of the equation $2x+3y=12$ are (6, 0) and (0, 4) respectively. The line of the equation $x=3$ is the vertical line parallel to y-axis, which intercepts the x-axis at (3, 0). The line $x=3$ intercepts the line $2x+3y=12$ at (3, 2).

The vertices of the quadrilateral formed by the solution set of the inequalities are (0, 0), (0, 4), (3, 2), and (3, 0). It is a trapezium. The lengths of the parallel sides are 4 and 2. The height of the trapezium is 3. Therefore, area of trapezium

$$= \frac{1}{2} \times (4 + 2)3 = 9$$

86.The perimeter of a rectangle is 100 cm. Find the length of its sides when its area is maximum.

A)45, 5 B)40, 10 C)35, 15 D)30, 20 E)25, 25

Answer: E

Let x be the length and y be the breadth of the rectangle.

We know that, 2(x+y)=100. Therefore, x+y=50. y=50-x.

Area of the rectangle = $x(50-x)=50x-x^2=625-(x-25)^2$

Maximum area of the rectangle is, when $(x-25)^2$ is zero.

It implies x=25 and y=25.

87.The total cost of producing 'x' machines is $f(x)=2000+100x-0.1x^2$. Find the average cost for producing 100 machines.

A)130 B)120 C)110 D)100 E)90

Answer: C

Average cost = f(x)/x (Total cost/number of machines produced)

$$average\ cost = \frac{2000}{x} + 100 - 0.1x$$

$$when\ x = 100, average\ cost = 20 + 100 - 10 = 110$$

88.If the price of sugar is increased by 15% find how much percent a householder must reduce her consumption of sugar so as not to increase the expenditure.

A)15% B)10% C)$11\frac{1}{23}\%$ D)$12\frac{1}{23}\%$ E)$13\frac{1}{23}\%$

Answer: E

Let the original price of sugar be $20.

The current cost of sugar is $23 (15% of 20 is 3).

The householder has only $20. Therefore, 3/23 part will be cut in the consumption.

$$\frac{3}{23} \times 100 = 13\frac{1}{23}\%$$

89.If the price of sugar is decreased by 15 percent find how much percent a householder can increase her consumption on the same budget.

A)$15\frac{11}{17}\%$ B)$16\frac{11}{17}\%$ C)$17\frac{11}{17}\%$ D)$18\frac{11}{17}\%$ E)$19\frac{11}{17}\%$

Answer: C

Let the original price of sugar be $20.

The current cost of sugar is $17 (15% of 20 is 3).

The householder now has $3 more. She can buy 3/17 part more.

$$\frac{3}{17} \times 100 = 17\frac{11}{17}\%$$

90.The value of a machine depreciates at the rate of 10% per annum. If its present value is $81,000, what was the value of the machine 2 years ago?

A)$98,010 B)$90,000 C)$99,000 D)$100,000 E)$99,010

Answer: D

If the value of the machine 1 year ago was 100, its present value is 90. As 90 correspond to $81,000, 100 correspond to $90,000.

$$\frac{81000}{90} \times 100 = 90000\ (value\ 1\ year\ ago)$$

If you do the same process once again, you will get the value of the machine 2 years ago.

$$\frac{90000}{90} \times 100 = 100{,}000$$

You can do this in a single stretch:

$$81000 \times \frac{100}{90} \times \frac{100}{90}$$

91.The average weight of a class of 24 students is 35 kilo gram. If the weight of the teacher is included, the average rises by 400 grams. Find the weight of the teacher.

A)45 kg B)44 kg C)43 kg D)42 kg E)41 kg

Answer: A

It is given that the average rises by 400 grams on the present strength of 25 (24 students and the teacher).

Therefore, the rise in total weight is 25x400=10,000 grams or 10 kg.

Hence, the teacher's weight is 35+10=45 kg.

92.A batsman makes a score of 87 runs in the 17th inning and thus increased his average by 3. Find the average after 17th inning.

A)36 B)37 C)38 D)39 E)40

Answer: D

The batsman's average increased by 3 after 17th inning. Therefore, the increase in total score will be 3x17=51.

Hence, his average before 17th inning is 87-51=36.

Therefore, his average after 17th inning is 36+3=39.

93.The average age of a committee of 8 members is 40 years. A member aged 55 years retired and a new member aged 39 joined in his place. Find the average age of the present committee.

A)37 B)38 C)39 D)40 E)41

Answer: B

The net decrease in the total age of the committee is 55-39=16.

Therefore, the decrease in average will be 16/8 =2.

Hence, the new average is 40-2=38.

94.A dishonest dealer professes to sell his goods at cost price but uses a weight of 960 grams for 1 kg. Find his gain percent.

A)4% B)$4\frac{1}{3}\%$ C)$4\frac{1}{4}\%$ D)$4\frac{1}{5}\%$ E)$4\frac{1}{6}\%$

Answer: E

The cost price now corresponds to 960 grams. Therefore, the dealer gains 40/960 part of the cost price. Hence, his gain percent is:

$$\frac{40}{960} \times 100 = 4\frac{1}{6}\%$$

95.750 men have provisions for 20 days. If at the end of 4 days, 450 men join the existing force, how long will the remaining provision last?

A)10 days B)11 days C)12 days D)13 days E)14 days

Answer: A

At the end of 4 days, 750 men have provisions for 16 days.

Now, there are 1200 men, who will share the same provision.

$$750 -> 16$$

$$1200 -> x$$

As number of men increases, number of days will decrease. They are inversely proportional. Therefore, use direct multiplication.

$$750 \times 16 = 1200 \times x$$

$$x = 10\ days.$$

96.If x men working x hours can produce x units in x days, then how many units can y men working y hours in y days produce?

A)x^3/y^3 B)x^3/y^2 C)y^3/x^3 D)y^3/x^2 E)y

Answer: D

$$x\,men \quad x\,hours \quad x\,days \quad x\,units$$

$$y\,men \quad y\,hours \quad y\,days \quad z\,units$$

We have to find z. Men, hours, and days, are directly proportional to number of units produced. Therefore, use cross multiplication.

$$z \times x \times x \times x = x \times y \times y \times y$$

$$z = \frac{y^3}{x^2}$$

97.A sum was put at simple interest at a certain rate for 2 years. Had it been put at 3% higher rate, it would have fetched $300 more. Find the sum.

A)$2,000 B)$2,500 C)$3,500 D)$4,500 E)$5,000

Answer: E

For a 3% higher rate, the sum fetched $300 more in 2 years.

Therefore, it would fetch $150 more in 1 year.

If 3% corresponds to $150, then 100% will correspond to:

$$\frac{150}{3} \times 100 = \$5{,}000$$

98.What annual installment will discharge a debt of $2,210 due in 4 years at 7% simple interest?

A)$450 B)$475 C)$500 D)$525 E)$550

Answer: C

$$M = An + \frac{Ar}{100} \times \frac{(n-1)n}{2}$$

$$M = amount\ due, A = annual\ installment, r = interest\ rate,$$

$$n = number\ of\ years$$

$$2210 = 4A + .07A \times \frac{3 \times 4}{2}$$

$$A = \frac{2210}{4.42} = \$500$$

99.The difference between compound interest and simple interest on a certain sum at 10% p.a. for 2 years is $52. Find the sum.

A)$5,200 B)$5,100 C)$5,000 D)$5,300 E)$5,400

Answer: A

Method 1

The difference is the interest amount on the interest amount of the first year. Therefore, the interest amount at the end of the first year is $520 (10% of 520=52).

Hence, the sum is $5,200 (10% of 5200=520).

Method 2

$$d = p(\frac{r}{100})^2$$

$$d = difference\ of\ SI\ and\ CI\ for\ 2\ years, p = pricipal, r = rate\ per\ annum$$

$$52 = p(0.1)^2$$

$$p = \$5{,}200$$

100.In what ratio must a person mix three kinds of wheat costing him \$1.20, \$1.44, and \$1.74 per kg so that the mixture may be worth \$1.41 per kg?

A)7:7:12 B)7:12:7 C)12:7:7 D)12:9:7 E)12:9:9

Answer: C

Use cheaper price-dearer price-mean price technique.

Take the first product and the second product.

Mean price-cheaper price: dearer price-mean price=second product: first product

1.41-1.20:1.44-1.41=21:3=7:1

Now, take the first product and the third product.

1.41-1.20:1.74-1.41=21:33=7:11

This is the ratio of third product to first product.

Add the first product in the two ratios. We get 1+11=12

Therefore, the ratio of first product : second product : third product is 12:7:7.

101.A and B solved a quadratic equation. In solving it, A made a mistake in the constant term and got the roots as 6 and 2, while B made a mistake in the coefficient of x only and got the roots as -7 and -1. Find the correct roots.

A)(7, 1) B)(2, 7) C)(6, -1) D)(6, -7) E)(2, -1)

Answer: A

The equation of A is $(x-6)(x-2)=x^2-8x+12$.

In this equation, the constant term 12 is wrong. However, the other terms are correct. Let the correct roots are α and β. $\alpha+\beta=8$; $\alpha\beta\neq 12$.

The equation of B is $(x+7)(x+1)=x^2+8x+7$.

In this equation, coefficient of x is wrong. However, the other terms are correct. Therefore, $\alpha+\beta\neq -8$; $\alpha\beta=7$.

Combining the two results, we get sum of the roots is 8 and product of the two roots is 7.

$$(\alpha-\beta)^2=(\alpha+\beta)^2-4\alpha\beta$$

$$(\alpha-\beta)^2=64-28=36$$

$$=>\alpha-\beta=\mp 6$$

Solving, we get two solution sets:(7, 1) or (1, 7).

Printed in the United States
By Bookmasters